Green Chemistry in Scientific Literature

Green Chemistry in Scientific Literature

A Bibliometric Study and Research Trends

Sanjay K. Sharma and Hasan Demir

CRC Press
Taylor & Francis Group
Boca Raton London New York

CRC Press is an imprint of the
Taylor & Francis Group, an **informa** business

CRC Press
Taylor & Francis Group
6000 Broken Sound Parkway NW, Suite 300
Boca Raton, FL 33487-2742

First issued in paperback 2022

ISBN-13: 978-0-367-43086-3 (hbk)
ISBN-13: 978-1-03-233758-6 (pbk)
DOI: 10.1201/9781003002352

Publisher's Note

The publisher has gone to great lengths to ensure the quality of this reprint but points out that some imperfections in the original copies may be apparent.

Visit the Taylor & Francis Web site at
http://www.taylorandfrancis.com

and the CRC Press Web site at
http://www.crcpress.com

Contents

List of Figures

List of Tables

Preface

GREEN CHEMISTRY, WHICH IS also known as "sustainable chemistry," enables chemists to create safe, energy-efficient, and non-toxic chemical products and processes, without harming human health and the environment. It is a more sophisticated way of doing chemistry, aiming at preventing pollution and health problems at the chemical design stage.

Green chemistry is more of a "chemistry FOR the environment," i.e., a more environmentally friendly chemistry rather than environmental chemistry, i.e., "chemistry OF the environment," that explains nature and the impact of man on the nature.

Green chemistry is based on a set of radical ideas (a set of 12 Scientific Principles) that overlap with the principles of sustainability and propose a modern version of chemistry that is less toxic, less hazardous, highly efficient, and non-polluting.

Being Book Series Editor of the "Green Chemistry for Sustainability" Book Series of Springer UK and associated as editorial board member and reviewer for many international research journals, Sanjay K. Sharma personally feels that the use of "green chemistry" and its popularization is very much required nowadays. Publication of such research in chemistry journals is a very useful source of spreading this awareness among students, research scholars, and faculty actively involved in research. And a systematic study of the present data will help a lot in this regard. In science and particularly in chemistry there is no such comprehensive bibliometric study available *so far*, which can give any idea about the share of "green chemistry" in chemistry journals. That's why he found this topic of proposal relevant and important. He discussed the idea with Dr. Hasan Demir, who readily convinced and agreed with him, and that's how this book was initiated.

The present book, entitled *Green Chemistry in Scientific Literature: A Bibliometric Study and Research Trends* consists of five comprehensive chapters. The study comprehends review and bibliometric features and characteristics of the research and review papers related to green chemistry. The bibliometric study was made by investigating literature indexing in Web of Science from 1999 to 2018. The bibliometrics analysis of 17,889 scientific papers (research and review papers) were processed according to year-wise and geographical-wise distribution, authorship pattern and citation, etc. The general information about green chemistry and policy of countries (US, Canada, and Europe) on green chemistry is mentioned. Twelve principles of green chemistry were also reviewed and analyzed by supporting statistical data. Research topics such as reaction media (ionic liquids and supercritical CO_2), green catalysis, polymers, photochemistry, microwave chemistry, and renewable resources are explained in the fourth chapter.

Acknowledgment

It is time to express our gratitude to our family, friends, colleagues, and well-wishers for extending their never-ending cooperation and support during the journey of this book.

We'd really like to thank Hilary Lafoe for providing us with the opportunity to write this book. We appreciate that she believed in us to make this book a reality.

We both thank every person of our parent institute and are grateful to work with such a beautiful group of people.

Sanjay K. Sharma is thankful to National Science and Technology Management Information System Division (NSTMIS), Department of Science and Technology, Government of India for financial support of this study.

Last, but not least, we are thankful to our students and readers, because they are the actual strength behind every teacher, every author, and every book. I am sure they will find this book useful.

Sanjay K. Sharma, FRSC

Hasan Demir

Authors

Prof. (Dr.) Sanjay K. Sharma is Fellow of Royal Society of Chemistry (UK) and a very well-known author and editor of many books, research journals, and hundreds of articles from last twenty years. He has also been appointed by Springer as Series Editor of the prestigious "Green Chemistry for Sustainability" series where he has been involved to date in the editing of 34 different titles of various international contributors.

Presently Prof. Sharma is Professor of Chemistry and Associate Dean (Research), JECRC University, Jaipur, India, where he teaches Engineering Chemistry and Environmental Chemistry to B. Tech. students and Green Chemistry, Spectroscopy and Organic Chemistry to M.S. students; pursues his research interest in the domain of Green Chemistry with special reference to water pollution, corrosion inhibition and biopolymers; and takes care of the Ph.D. program as well as research activities of the University.

Dr. Sharma has 19 books of chemistry from national/international publishers and over 100 research papers of national and international repute to his credit.

He is also a member of American Chemical Society (USA), and International Society for Environmental Information Sciences (ISEIS, Canada), and is also a life member of various international professional societies including International Society of Analytical Scientists, Indian Council of Chemists, International Congress of Chemistry and Environment, Indian Chemical Society.

Dr. Sharma also serves as Editor-in-Chief for *RASAYAN Journal of Chemistry* (A SCOPUS, Elsevier, indexed Journal) and is also associated as editorial member for more than 20 other international research journals.

He provides resource services in areas of green chemistry, water sustainability, manuscript writing and scientific ethics for various prestigious national and international organizations.

Hasan Demir has a bachelor's degree in chemical engineering from Ege University in Turkey. He received a Ph.D. from Izmir Institute of Technology in 2009 for work concerning adsorption heat pumps. Additionally, he has a post-doctoral study from Technion Institute of Technology in Israel in the Department of Energy, Faculty of Mechanical Engineering. He has been an Associate Professor at Osmaniye Korkut Ata University in Turkey since 2010. He finished two TUBITAK (The Scientific and Technological Research Council of Turkey) projects, entitled "Effect of Microwave Regenerated Adsorbent Bed on The Performance of Adsorption Heat Pump" and "Optimization and Improvement of the Design of an Open Liquid Desiccant System." He has published 22 papers which have received more than 900 citations, three book chapters, and 20 articles in conferences. His recent project (318O122) concerns the fabrication of commercial adsorption chiller with an industrial partner.

CHAPTER 1

Introduction to Green Chemistry and Policy

1.1 GENERAL INFORMATION

The U.S. Environmental Protection Agency (EPA) defines green chemistry as the design of chemical products and processes that reduce or eliminate the generation of hazardous substances. The concept of green chemistry involves renewable raw materials, eliminating waste, and avoiding hazardous use of toxic reagents and solvents in the chemical industry (Gupta and Paul, 2014). Anastas and Warner (Anastas, 1998) have stated 12 principles that draw the frame of green chemistry definition. These principles are

- Preventing waste generation
- Increasing conversion of reactions
- Designing less hazardous chemical syntheses
- Designing safer chemicals and products
- Using safer solvents and reaction conditions
- Increasing energy efficiency
- Using renewable feedstocks

- Avoiding chemical derivatives
- Using catalysts rather than stoichiometric reagents
- Designing chemicals and products to degrade after use
- Analyzing in real-time to prevent pollution
- Minimizing the potential for accidents (EPA, 2018)

A variety of research areas related to green chemistry were comprehensively reviewed by Erythropel et al. (2018).

Catalysis, which is widely used in the chemical industry, is one of the ways of implementing the principles of green chemistry (Gupta and Paul, 2014). Some greener chemicals can be used as a catalyst in Friedel–Crafts alkylations reaction (Rueping and Nachtsheim, 2010), Mannich reaction (Iwanejko et al., 2018), synthesis of fine chemical (dioxygen activation) (Romero-Guido et al., 2018), Suzuki cross-coupling reaction (Franzén and Xu, 2005), nano-catalysis in aqueous reactions (Polshettiwar and Varma, 2010), photocatalytic reactions (Michelin and Hoffmann, 2018)(König, 2017), catalytic nanoreactors to carry out organic reactions (De Martino et al., 2018), green polymer synthesis and production (Marszałek-Harych et al., 2017) (Dubé and Salehpour, 2014) (Anderson et al., 2018), organocatalytic oxidative reactions (Triandafillidi et al., 2018), macromolecular (i.e., polysaccharides) synthesis (Shoda et al., 2016), heterogeneous catalytic oxidations with hydrogen peroxide, homogeneous catalytic oxidations and carbonylations and organocatalytic oxidations with stable N-oxy radicals (Sheldon, 2008), acethylene hydrochlorination reactions (Xu and Luo, 2018), application of Lacase in oxidation reactions (Witayakran and Ragauskas, 2009), and oxidative catalysis in aqueous environment (Ryabov, 2013). The green chemicals were produced and/or utilized in some processes such as synthesis of three- to five-membered O-heterocycles (Kaur, 2018), synthesis of acyclic carbamates (Hosseinian et al., 2018), synthesis enzymatic polymer (Kobayashi and Makino, 2009), synthesis of metal-organic frameworks (Reinsch, 2016), synthesis of metal nanoparticles (Vaseghi et al., 2018), applications through metagenomic technology (Castilla et al., 2018), development of ionic liquid gel materials (Marr and Marr, 2015), and heterocyclic compounds synthesis (M'Hamed, 2015).

The chemical industry uses multiple solvents in different steps of a process, which has potential to reduce waste and environmental impacts (Dunn, 2012) (Erythropel et al., 2018) (Mulvihill et al., 2011). Vegetable oils (Li et al., 2017) and glycerol (Gu and Jérôme, 2010) are used as a solvent to extract compounds for purification, enrichment, or even pollution remediation. Green techniques which are pulsed electric field, microwave (Christopher R. Strauss, 2015), instantaneous controlled pressure drop, application of ultrasound (Cintas and Luche, 1999), etc., is used for solvent-free extraction of food and natural products (i.e., olive oil and citrus essential oil) (Armenta, 2015) (Chemat et al., 2015). However, the concept of green chemistry is not new in the pharmaceutical industry; there are some significant efforts to increase efficiency, quality, control, and waste reduction (Gupta and Mahajan, 2015). Membrane extraction techniques are one of the new and green techniques with special focus on pharmaceutical and biomedical analysis since 2013 (Tabani et al., 2018). Nanotechnology is an interdisciplinary branch of science involving chemistry, physics, biology, engineering, and, recently, toxicology. Nanotechnology also appears in green chemistry and the pharmaceutical industry in the context of potential health effects of nanoparticles, along with medical applications of nanoparticles (i.e., cellulose nanofibers (Isogai and Bergström, 2018), metal nanoparticles (Vaseghi et al., 2018) (Makarov et al., 2014), selenium nanoparticles (Shoeibi et al., 2017), carbon nanomaterials (Basiuk and Basiuk, 2014), and gold nanoparticles (Chen and Liu, 2018)), including visualization, drug delivery, disinfection, water cleaning, desalination, and tissue repair (Villaseñor and Ríos, 2018; Peel and Roberts, 2003; Duan et al., 2015; Ghernaout et al., 2011). New approaches in the pharmaceutical industry cause improvement of efficiency in quality and productivity, as well as a decrease in toxic/hazardous solvent consumption (Galyan et al., 2018). Chromatographic techniques have significant impact on the pharmaceutical industry and the potential to be greener at all steps of the analysis, and they can be integrated into workflows to increase productivity in drug production and purification processes (Galyan et al., 2018) (Korany et al., 2017) (Justyna Małgorzata Płotka-Wasylka et al., 2018) .

In the last two decades, the chemical and process industries have appeared to transform from traditional disciplines to sustainable products, processes, and production systems (García-Serna et al., 2007) (Melchert et al., 2012). Garcia-Serna et al. (2007) reviewed the illustrative industrial applications and case studies in green engineering, including frameworks for design and legislative aspects. Some of the new trends in liquid-phase chemical processes are: using ionic liquids (Kitazume, 2000), sub- or supercritical fluids, and some neoteric solvents that have gained considerable attention for preparation and modification of important chemical compounds and materials (Kuchurov et al., 2017). Kuchurov et al. (2017) reviewed the significant potential of these compounds on nitration and other reactions used for manufacturing and processing of high-energy materials (HEMs). Processing of lignin, an important source of renewable and bio-based carbon, is summarized and current routes in accordance with green chemistry principles are explained by Gillet et al. (2017). A microelectronic sector, one of the largest manufacturing sectors in the world, investigates the alternative greener process technologies. O'Neil and Watkins (2004) examine a review of the use of supercritical fluids such as carbon dioxide in advanced device fabrication. Nacca et al., (2017) reviewed applications of Q-tube, which are recently developed and cheap, easy to handle, safe, and highly versatile devices in organic synthesis (Nacca et al., 2017). In the textile industry, various biopolymers are developed for antimicrobial textiles by taking sustainability, environmental friendliness, reduced pollution, and green chemistry into consideration (Shahid-Ul-Islam et al., 2013). Biofuels, biomass, and biorefinery technology assist to become widespread green, with the potential of utilizing waste as a new resource, the transformation of available facilities to biorefinery, and producing multiple products from biomass (Clark et al., 2012) (Gude and Martinez-Guerra, 2018) (Ryan and Senge, 2015) (Van Schoubroeck et al., 2018). Dichiarante et al. (2010) reported the survey of literature on green chemistry, mainly concerning more environment-friendly synthetic methods, catalytic systems, less harmful solvents, and alternative physical techniques (Dichiarante et al., 2010).

1.2 POLICY

The chemical production rate is €1,244 billion, with 31% for the EU chemical industry and with 28% of the production value for the US in 1998 (EC, 2001). Laws and regulations assist the sectors to meet the principles of green chemistry. The United States EPA revealed some regulations to help sectors prevent pollution and protect the environment. The mentioned sectors are: agriculture, automotive, construction, electric utilities, oil and gas, transportation, mining, and manufacturing. (EPA, 2018). The process of pesticide registration in agriculture includes an examination of the ingredients of pesticide, the particular site or crop where it is to be used, the amount, frequency, and timing of its use, and storage and disposal practices. In evaluating the process of registration of a pesticide, the risk assessment takes into account harm to humans, wildlife, fish, plants, and contamination of surface water or groundwater. In the United States, laws on all varieties of pesticides are gathered into two federal regulations, which are the Federal Insecticide, Fungicide, and Rodenticide Act (FIFRA) and the Federal Food, Drug, and Cosmetic Act (FFDCA) (EPA, 2018). The distribution, sale, and use of pesticide is regulated by the FIFRA 7 U.S.C. §136 et seq. (1996). The maximum tolerances of pesticide residues on foods are set by the FFDCA 21 U.S.C. §301 et seq. (2002). More regulations and laws on the environment by different sectors can be found from the EPA website. Some of the examples for the strict enforcement of EPA regulations in the fiscal year of 2018 was accomplished as follows:

- Prevent the importing of approximately 2,200 vehicles and engines which fail to abide with EPA standards.
- Reduce exposure to lead through 140 enforcement actions.
- Invest approximately $4 billion in actions and equipment for the agreement of EPA standards.
- A total of 73 years of imprisonment for the criminal found guilty.

Figure 1.1 illustrates the benefits of EPA regulations on the commitment to reduce pollution of air and water in pounds per year. As seen in the following figure, environmental pollution tends to gradually reduce after 2013. In 2018, environmental pollution is observed as 268M pounds. Briefly, enforcement of EPA regulation causes a decrease in environmental pollution.

The Canadian Environmental Protection Act (CEPA, 1999) aspires assessment and management of chemicals for the

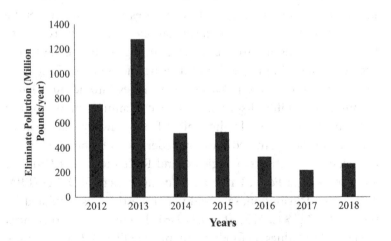

FIGURE 1.1 The commitment to reduce pollution of air and water between 2012 and 2018.

protection of the environment and human health. The CEPA describes a toxic level of chemical substance according to the following features ("CEPA, 1999," 2019):

- Have an adverse effect on the environment

- Are damaging to the environment that life depends on

- Could be damaging to human health or life.

Compliance of people with CEPA can achieve desired environmental and living being (plant, human, and animal) health. Incremental awareness, understanding, and compliance with the regulations causes a decrease in environmental damage and enforcement actions. The CEPA supports enforcement actions which are prohibition, recall and detention orders for ships, environmental protection compliance orders and environmental violations administrative monetary penalties, etc. ("CEPA, 1999," 2019). The CEPA makes planned and unplanned inspections and investigations to control disagreement of environmental regulations activities. Figure 1.2 illustrates the number of inspections under CEPA from April 2016 to March 2017.

The controls and management of chemical policy in the EU are developed in 1999. In 2001, the European Commission released a white paper called "strategy for a future chemical policy." In 1981, a defined number of substances in the market was 100,106; with the white paper, the number of existing chemical substances in the market in volumes above 1 tonne was estimated at 30,000. Of the existing chemical substances, 140 have been defined as priority chemicals for risk assessment and management (EC, 2001) (CELA, 2007). The chemical policy of EU is more effective and efficient and obligations are extended with new regulations called Registration, Evaluation and Authorization of Chemicals (REACH) in 2006 (EU Council, 2006). All chemical substances above 1 tonne should be registered and fulfill the requirements involving testing requirements. The restrictions on classification,

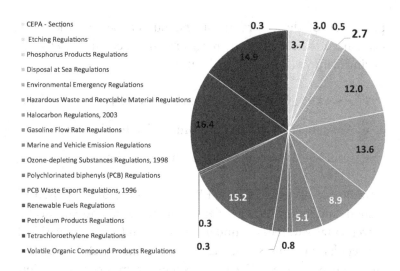

CEPA - Sections
Etching Regulations
Phosphorus Products Regulations
Disposal at Sea Regulations
Environmental Emergency Regulations
Hazardous Waste and Recyclable Material Regulations
Halocarbon Regulations, 2003
Gasoline Flow Rate Regulations
Marine and Vehicle Emission Regulations
Ozone-depleting Substances Regulations, 1998
Polychlorinated biphenyls (PCB) Regulations
PCB Waste Export Regulations, 1996
Renewable Fuels Regulations
Petroleum Products Regulations
Tetrachloroethylene Regulations
Volatile Organic Compound Products Regulations

FIGURE 1.2   The number of inspections under CEPA from April 2016 to March 2017.

packaging, labeling, marketing, and use of certain dangerous substances are defined by directives such as (76/769/(EEC, 1976)) and (67/548/(EEC, 1967)). In order to provide sustainable environmental pollution control, the Organization for Economic Co-operation and Development (OECD) defined "extended producer (some case production) responsibility, EPR" in 2003 as

> An environmental policy approach in which a producer's responsibility, physical and/or financial, for a product is extended to the post-consumer stage of a product's life cycle. There are two related features of EPR policy: (1) the shifting of responsibility (physically and/or economically; fully or partially) upstream to the producer and away from municipalities, and (2) to provide incentives to producers to incorporate environmental considerations in the design of their products. (OECD, 2001).

An example of environmental policy on packaging waste, which is an important issue in modern life (as is the growth of consumption and transportation), can be given as follows for EU communities, Canada, and the United States. The recycling rate of packaging materials in the United States and Canada (Ontario) were 39% and 41.8% in 2003, respectively (CELA, 2007). Figure 1.3 illustrates the percentage and the total amount of waste, recovery, and recycling of domestic packaging produced in 28 EU communities from 2007 to 2016. The total amount of packaging gradually increases after 2013. In general, approximately 60% of the produced total domestic packaging is recycled or recovered. Only 40% of total domestic packaging is subjected to waste generation (Eurostat, 2019a).

The weight ratio of types of packaging for waste generated in EU communities is presented in Figure 1.4. The maximum weight ratio of packaging waste generated was approximately 40% and belongs to paper and cardboard packaging materials. The weight ratios of plastic and wooden packaging as waste generated were 20% and 15%, respectively. The weight ratios of glass packaging as waste generated were 20% (Eurostat, 2019a). Variation of weight ratios of types of packaging does not change for recycling and recovery packaging. In other words, the weight ratio of types of packaging for waste generated is similar for recycling and recovery packaging.

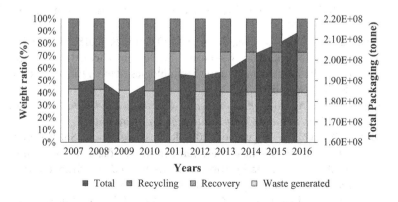

FIGURE 1.3   Percentage and the total amount of waste generated and recovery and recycling of domestic packaging for 28 EU communities.

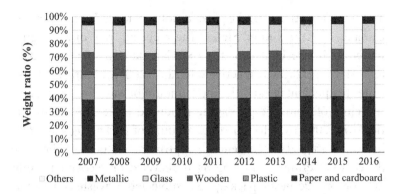

FIGURE 1.4 The weight ratio of types of packaging for waste generated in 28 EU communities.

REFERENCES

Anastas, P.T., Warner, J.C., 1998. *Green Chemistry: Theory and Practice.* Oxford University Press.

Anderson, L.A., Islam, M.A., Prather, K.L.J., 2018. Synthetic biology strategies for improving microbial synthesis of "green" biopolymers. *J. Biol. Chem.* 293, 5053–5061. doi:10.1074/jbc.TM117.000368

Armenta, S., Garrigues, S., de la Guardia, M., 2015. The role of green extraction techniques in Green Analytical Chemistry. *Trends Anal. Chem.* 71, 2–8. doi:10.1016/j.trac.2014.12.011

Basiuk, E.V., Basiuk, V.A., 2014. Green chemistry of carbon nanomaterials. *J. Nanosci. Nanotechnol.* 14, 644–672. doi:10.1166/jnn.2014.9011

Castilla, I.A., Woods, D.F., Reen, F.J., O'Gara, F., 2018. Harnessing marine biocatalytic reservoirs for green chemistry applications through metagenomic technologies. *Mar. Drugs* 16, 1–21. doi:10.3390/md16070227

CELA, C.E.L.A., 2007. *European and Canadian Environmental Law : Best Practices and Opportunities for Co-operation.*

CEPA 1999 [WWW Document], 2019. Risk Assess. Chem. Subst. CEPA. URL https://www.canada.ca/en/health-canada/services/chemical-substances/canada-approach-chemicals/canadian-environmental-protection-act-1999.html (accessed 3.5.19).

Chemat, F., Fabiano-Tixier, A.S., Vian, M.A., Allaf, T., Vorobiev, E., 2015. Solvent-free extraction of food and natural products. *TrAC - Trends Anal. Chem.* 71, 157–168. doi:10.1016/j.trac.2015.02.021

Chen, H.C., Liu, Y.C., 2018. Creating functional water by treating excited gold nanoparticles for the applications of green chemistry, energy and medicine: A review. *J. Ind. Eng. Chem.* 60, 9–18. doi:10.1016/j.jiec.2017.09.026

Christopher, R. Strauss, R.S.V., 2015. Microwaves in green and sustainable chemistry. *Handb. Environ. Chem.* 32, 405–428. doi:10.1007/698_2014_267

Cintas, P., Luche, J.L., 1999. Green chemistry: The sonochemical approach. *Green Chem.* 1, 115–125. doi:10.1039/a900593e

Clark, J.H., Luque, R., Matharu, A.S., 2012. Green chemistry, biofuels, and biorefinery. *Annu. Rev. Chem. Biomol. Eng.* 3, 183–207. doi:10.1146/annurev-chembioeng-062011-081014

Council, E., 2006. Regulation (EC) No 1907/2006. *Off. J. Eur. Union.*

De Martino, M.T., Abdelmohsen, L.K.E.A., Rutjes, F.P.J.T., Van Hest, J.C.M., 2018. Nanoreactors for green catalysis. *Beilstein J. Org. Chem.* 14, 716–733. doi:10.3762/bjoc.14.61

Dichiarante, V., Ravelli, D., Albini, A., 2010. Green chemistry: State of the art through an analysis of the literature. *Green Chem. Lett. Rev.* 3, 105–113. doi:10.1080/17518250903583698

Duan, H., Wang, D., Li, Y., 2015. Green chemistry for nanoparticle synthesis. *Chem. Soc. Rev.* 44, 5778–5792. doi:10.1039/c4cs00363b

Dubé, M.A., Salehpour, S., 2014. Applying the principles of green chemistry to polymer production technology. *Macromol. React. Eng.* 8, 7–28. doi:10.1002/mren.201300103

Dunn, P.J., 2012. The importance of green chemistry in process research and development. *Chem. Soc. Rev.* 41, 1452–1461. doi:10.1039/c1cs15041c

EC, 2001. *Strategy for a Future Chemicals Policy.*

EEC, 1967. *European Economic Community, and in Particular.* European Community.

EEC, 1976. *Having Regard to the Treaty Establishing the European.* European Community.

EPA, 2018. US Environmental Protection Agency [WWW Document]. web page. URL https://www.epa.gov/greenchemistry/basics-gre en-chemistry#definition (accessed 11.21.18).

Erythropel C.H.C., Zimmerman J., de Winter T., Petitjean L., Melnikov F., Lam C.H., Lounsbury A., Mellor K., Janković N., Tu Q., Pincus L.N., 2018. The Green ChemisTREE: 20 years after taking root with the 12 Principles. *Green Chem* 20(9): 1929–1961. doi:10.1039/v8gc00482j.

Eurostat, 2019. Eurostat: Statistics explained [WWW Document]. Packag. waste Stat. URL https://ec.europa.eu/eurostat/statistics

-explained/index.php/Packaging_waste_statistics (accessed 3.8.19).

Franzén, R., Xu, Y., 2005. Review on green chemistry—Suzuki cross coupling in aqueous media. *Can. J. Chem.* 83, 266–272. doi:10.1139/v05-048

Galyan, K., Reilly, J., Galyan, K., Reilly, J., 2018. Green chemistry approaches for the purification of pharmaceuticals. *Curr. Opin. Green Sustain. Chem.* doi:10.1016/j.cogsc.2018.04.018

García-Serna, J., Pérez-Barrigón, L., Cocero, M.J., 2007. New trends for design towards sustainability in chemical engineering: Green engineering. *Chem. Eng. J.* 133, 7–30. doi:10.1016/j.cej.2007.02.028

Ghernaout, D., Ghernaout, B., Naceur, M.W., 2011. Embodying the chemical water treatment in the green chemistry-A review. *Desalination* 271, 1–10. doi:10.1016/j.desal.2011.01.032

Gillet, S., Aguedo, M., Petitjean, L., Morais, A.R.C., Da Costa Lopes, A.M., Łukasik, R.M., Anastas, P.T., 2017. Lignin transformations for high value applications: Towards targeted modifications using green chemistry. *Green Chem.* 19, 4200–4233. doi:10.1039/c7gc01479a

Gu, Y., Jérôme, F., 2010. Glycerol as a sustainable solvent for green chemistry. *Green Chem.* 12, 1127–1138. doi:10.1039/c001628d

Gude, V.G., Martinez-Guerra, E., 2018. Green chemistry with process intensification for sustainable biodiesel production. *Environ. Chem. Lett.* 16, 327–341. doi:10.1007/s10311-017-0680-9

Gupta, P., Mahajan, A., 2015. Green chemistry approaches as sustainable alternatives to conventional strategies in the pharmaceutical industry. *RSC Adv.* 5, 26686–26705. doi:10.1039/C5RA00358J

Gupta, P., Paul, S., 2014. Solid acids: Green alternatives for acid catalysis. *Catal. Today* 236, 153–170. doi:10.1016/j.cattod.2014.04.010

Hosseinian, A., Ahmadi, S., Mohammadi, R., Monfared, A., Rahmani, Z., 2018. Three-component reaction of amines, epoxides, and carbon dioxide: A straightforward route to organic carbamates. *J. CO2 Util.* 27, 381–389. doi:10.1016/j.jcou.2018.08.013

Isogai, A., Bergström, L., 2018. Preparation of cellulose nanofibers using green and sustainable chemistry. *Curr. Opin. Green Sustain. Chem.* 12, 15–21. doi:10.1016/j.cogsc.2018.04.008

Iwanejko, J., Wojaczyńska, E., Olszewski, T.K., 2018. Green chemistry and catalysis in Mannich reaction. *Curr. Opin. Green Sustain. Chem.* 10, 27–34. doi:10.1016/j.cogsc.2018.02.006

Kaur, N., 2018. Green synthesis of three- to five-membered O-heterocycles using ionic liquids. *Synth. Commun.* 48, 1588–1613. doi:10.1080/00397911.2018.1458243

Kitazume, T., 2000. Green chemistry development in fluorine science. *J. Fluor. Chem.* 105, 265–278. doi:10.1016/S0022-1139(99)00 269-9

Kobayashi, S., Makino, A., 2009. Enzymatic polymer synthesis: An opportunity for green polymer chemistry. *Chem. Rev.* 109, 5288–5353. doi:10.1021/cr900165z

Korany, M.A., Mahgoub, H., Haggag, R.S., Ragab, M.A.A., Elmallah, O.A., 2017. Green chemistry : Analytical and chromatography. *J. Liq. Chrom. Rel. Technol.* 40, 839–852. doi:10.1080/10826076.201 7.1373672

König, B., 2017. Photocatalysis in organic synthesis – past, present, and future. *European J. Org. Chem.* 2017, 1979–1981. doi:10.1002/ ejoc.201700420

Kuchurov, I.V., Zharkov, M.N., Fershtat, L.L., Makhova, N.N., Zlotin, S.G., 2017. Prospective symbiosis of green chemistry and energetic materials. *Chem. Sus. Chem.* 10, 3914–3946. doi:10.1002/cssc. 201701053

Li, Y., Fabiano-Tixier, A.S., Chemat, F., 2017. Vegetable oils as alternative solvents for green extraction of natural products. *Edible Oils Extr. Process. Appl.* 205–222. doi:10.1201/9781315152493

M'Hamed, M.O., 2015. Ball milling for heterocyclic compounds synthesis in green chemistry: A review. *Synth. Commun.* 45, 2511–2528. doi:10.1080/00397911.2015.1058396

Makarov, V.V., Love, A.J., Sinitsyna, O.V., Makarova, S.S., Yaminsky, I.V., Taliansky, M.E., Kalinina, N.O., 2014. "Green" nanotechnologies: Synthesis of metal nanoparticles using plants. *Acta Naturae* 6, 35–44. doi:10.1039/c1gc15386b

Małgorzata Płotka-Wasylka Justyna, Susdorf, A.K., Sajid Muhammad, de la Guardia Miguel, Namieśnik, J., and Tobiszewski M., 2018. Green chemistry in higher education: State of the art, challenges and future trends. *Chem. Sustain.* doi:10.1002/cssc.201801109

Marr, P.C., Marr, A.C., 2015. Ionic liquid gel materials: Applications in green and sustainable chemistry. *Green Chem.* 18, 105–128. doi:10.1039/c5gc02277k

Marszałek-Harych, A., Jedrzkiewicz, D., Ejfler, J., 2017. Bio- and chemocatalysis cascades as a bridge between biology and chemistry for green polymer synthesis. *Cell. Mol. Biol. Lett.* 22, 1–14. doi:10.1186/ s11658-017-0061-1

Melchert, W.R., Reis, B.F., Rocha, F.R.P., 2012. Green chemistry and the evolution of flow analysis. A review. *Anal. Chim. Acta* 714, 8–19. doi:10.1016/j.aca.2011.11.044

Michelin, C., Hoffmann, N., 2018. Photocatalysis applied to organic synthesis – A green chemistry approach. *Curr. Opin. Green Sustain. Chem.* 10, 40–45. doi:10.1016/j.cogsc.2018.02.009

Mulvihill, M.J., Beach, E.S., Zimmerman, J.B., Anastas, P.T., 2011. Green chemistry and green engineering : A framework for sustainable technology development. *Ann. Rev. Environ. Res.* 36, 271–293. doi:10.1146/annurev-environ-032009-095500

Nacca, F.G., Merlino, O., Mangiavacchi, F., Krasowska, D., Santi, C., Sancineto, L., 2017. The Q-tube system, a nonconventional technology for green chemistry practitioners. *Curr. Green Chem.* 4, 58–66. doi:10.2174/2213346104666170908160741

O'Neil, A., Watkins, J.J., 2004. Green chemistry in the microelectronics industry. *Green Chem.* 6, 363–368. doi:10.1039/b403729d

OECD, 2001. *Extended Producer Responsibility A Guidance Manual for Governments*. OECD Publisher, Paris.

Peel, M.J., Roberts, R., 2003. Audit fee determinants and auditor premiums: Evidence from the micro-firm sub-market. *Account. Bus. Res.* 33, 207–233. doi:10.1039/b517131h

Polshettiwar, V., Varma, R.S., 2010. Green chemistry by nano-catalysis. *Green Chem.* 12, 743–754. doi:10.1039/b921171c

Reinsch, H., 2016. "Green" synthesis of metal-organic frameworks. *Eur. J. Inorg. Chem.* 2016, 4290–4299. doi:10.1002/ejic.201600286

Romero-Guido, C., Baez, A., Torres, E., 2018. Dioxygen activation by laccases: Green chemistry for fine chemical synthesis. *Catalysts* 8, 223. doi:10.3390/catal8060223

Rueping, M., Nachtsheim, B.J., 2010. A review of new developments in the Friedel–Crafts alkylation – From green chemistry to asymmetric catalysis. *Beilstein J. Org. Chem.* 24, 1–24. doi:10.3762/bjoc.6.6

Ryabov, A.D., 2013. Green challenges of catalysis via Iron(IV)oxo and Iron(V)oxo species, in: *Advances in Inorganic Chemistry*. pp. 117–163. doi:10.1016/B978-0-12-404582-8.00004-3

Ryan, A.A., Senge, M.O., 2015. How green is green chemistry? Chlorophylls as a bioresource from biorefineries and their commercial potential in medicine and photovoltaics. *Photochem. Photobiol. Sci.* 14, 638–660. doi:10.1039/c4pp00435c

Shahid-Ul-Islam, Shahid, M., Mohammad, F., 2013. Green chemistry approaches to develop antimicrobial textiles based on sustainable biopolymers - A review. *Ind. Eng. Chem. Res.* 52, 5245–5260. doi:10.1021/ie303627x

Sheldon, R.A., 2008. E factors, green chemistry and catalysis: An odyssey. *Chem. Commun.* 3352–3365. doi:10.1039/b803584a

Shoda, S.I., Uyama, H., Kadokawa, J.I., Kimura, S., Kobayashi, S., 2016. Enzymes as green catalysts for precision macromolecular synthesis. *Chem. Rev.* 116, 2307–2413. doi:10.1021/acs.chemrev.5b00472

Shoeibi, S., Mozdziak, P., Golkar-Narenji, A., 2017. Biogenesis of selenium nanoparticles using green chemistry. *Top. Curr. Chem.* 375. doi:10.1007/s41061-017-0176-x

Tabani, H., Nojavan, S., Alexovič, M., Sabo, J., 2018. Recent developments in green membrane-based extraction techniques for pharmaceutical and biomedical analysis. *J. Pharm. Biomed. Anal.* 160, 244–267. doi:10.1016/j.jpba.2018.08.002

Triandafillidi, I., Tzaras, D.I., Kokotos, C.G., 2018. Green organocatalytic oxidative methods using activated ketones. *Chem. Cat. Chem.* 10, 2521–2535. doi:10.1002/cctc.201800013

Van Schoubroeck, S., Van Dael, M., Van Passel, S., Malina, R., 2018. A review of sustainability indicators for biobased chemicals. *Renew. Sustain. Energy Rev.* 94, 115–126. doi:10.1016/j.rser.2018.06.007

Vaseghi, Z., Nematollahzadeh, A., Tavakoli, O., 2018. Green methods for the synthesis of metal nanoparticles using biogenic reducing agents: A review. *Rev. Chem. Eng.* 34, 529–559. doi:10.1515/revce-2017-0005

Villaseñor, M., Ríos, Á., 2018. Nanomaterials for water cleaning and desalination, energy production, disinfection, agriculture and green chemistry. *Environ. Chem. Lett.* 16, 11–34. doi:10.1007/s10311-017-0656-9

Witayakran, S., Ragauskas, A.J., 2009. Synthetic applications of laccase in green chemistry. *Adv. Synth. Catal.* 351, 1187–1209. doi:10.1002/adsc.200800775

Xu, H., Luo, G., 2018. Green production of PVC from laboratory to industrialization: State-of-the-art review of heterogeneous non-mercury catalysts for acetylene hydrochlorination. *J. Ind. Eng. Chem.* 65, 13–25. doi:10.1016/j.jiec.2018.05.009

CHAPTER 2

Bibliometric Study

2.1 METHODOLOGY

The research and review papers were surveyed in the core collection of Web of Science from 1999 to 2018. The bibliometric analysis of 17,889 scientific papers were processed according to year-wise and geographical-wise distribution, authorship pattern and citation, etc. All these data were organized, analyzed, tabulated, and presented by using Microsoft Excel. The applied query was: (TS=(Green AND Chemistry)) AND DOCUMENT TYPES: (Article OR Review) AND PUBYEAR > 1998 AND PUBYEAR < 2019). The search was made on the 27th of June 2019.

2.2 SCIENTIFIC CATEGORIES

The 17,889 scientific papers are composed of 89.4% research articles and 10.6% review papers. Figure 2.1 presents the top 20 of the diversity of research categories in a scientific manner. 70% of the research was published under the subject category of "Chemistry." The "science technology" topic was observed with 12.8% of published papers. The "engineering," "material science," and "biochemistry molecular biology" categories involved 10.1%, 8.6%, and 5.6% of the number of published papers.

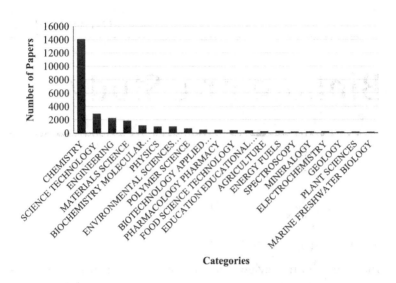

FIGURE 2.1   Top 20 scientific categories of published papers.

2.3 YEAR-WISE DISTRIBUTION OF ARTICLES

Figure 2.2 illustrates the year-wise distribution of published research and review papers. During the years between 1999 and 2018, the concepts of green chemistry have acquired growing attention. In 2018, 2,239 research and review papers were published

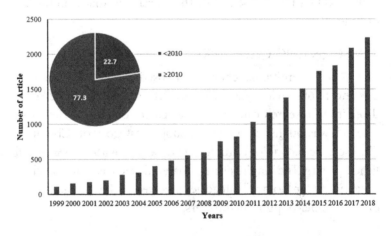

FIGURE 2.2   Year-wise distribution of the number of articles.

in journals indexed in *Science Citation Index Expanded*, *Journal Citation Reports*, and *Engineering Index*. The number of published papers was only 108 in 1999. From 1999 to 2018, the number of published papers increased approximately 20 times. From a different point of view, the pie chart in Figure 2.2 shows that 22.7% of the papers were published before 2010 and 77.3% of the papers were issued after 2010. The results reveal that concern about the concept of green chemistry has increased rapidly.

2.4 GEOGRAPHICAL-WISE DISTRIBUTION OF CONTRIBUTIONS

Table 2.1 presents the number of contributions to the green chemistry papers according to countries. Ninety-six countries contributed to the research and review papers. The total number of contributions is 114,415, involving national and international collaborations. For that reason, data given in this section does not present the number of papers that indicates the number of contributions of collaborative countries. The United States has made the most contribution to the articles with 22,718. China has made 17,203 contributions and Japan has made 7,038 contributions. India and Germany are the other countries in the top five and have made 7,052 and 6,292 contributions, respectively. The total number of contributions also indicates and emphasizes that green chemistry is an interdisciplinary topic, as can also be seen in the results of the following sections.

Figure 2.3 reveals the distribution of the number of contributions over continents. The highest contribution was from countries in Asia, with 38.3%. Europe contributed 31.5%. North America contributed 23.2%.

2.5 AUTHORSHIP PATTERN

Authorship pattern of contributions is presented in Figure 2.4. The single-authored paper has a minimum percentage, with 5.5%; it can be interpreted that the authors tend to research in collaboration. The maximum percentage was observed as 25.9% (4,626

TABLE 2.1 Country-Wise Distribution of Contributions

No	Country	Number of Papers	No	Country	Number of Papers	No	Country	Number of Papers
1	Algeria	120	33	Ghana	23	65	Poland	683
2	Argentina	553	34	Greece	473	66	Portugal	857
3	Australia	2059	35	Hungary	398	67	Qatar	30
4	Austria	716	36	India	7052	68	Romania	257
5	Bangladesh	66	37	Indonesia	149	69	Russia	1061
6	Belarus	52	38	Iran	2538	70	Saudi Arabia	715
7	Belgium	1077	39	Iraq	77	71	Serbia	132
8	Benin	13	40	Ireland	328	72	Singapore	735
9	Botswana	15	41	Israel	666	73	Slovakia	142
10	Brazil	2400	42	Italy	3172	74	Slovenia	182
11	Brunei	16	43	Japan	7038	75	South Africa	557
12	Bulgaria	134	44	Jordan	58	76	South Korea	2905
13	Cameroon	25	45	Kenya	91	77	Spain	3102
14	Canada	3029	46	Kuwait	42	78	Sri Lanka	55
15	Chad	7	47	Latvia	34	79	Sweden	1114
16	Chile	223	48	Lebanon	26	80	Switzerland	1335
17	China	17203	49	Lithuania	92	81	Taiwan	1345
18	Colombia	146	50	Luxembourg	48	82	Tanzania	27

(Continued)

TABLE 2.1 (CONTINUED) Country-Wise Distribution of Contributions

No	Country	Number of Papers	No	Country	Number of Papers	No	Country	Number of Papers
19	Costa Rica	37	51	Malaysia	713	83	Thailand	598
20	Croatia	108	52	Malta	1	84	Tunisia	142
21	Cuba	45	53	Mauritius	16	85	Turkey	803
22	Cyprus	27	54	Mexico	719	86	Uganda	20
23	Czechia	556	55	Morocco	96	87	Ukraine	123
24	Denmark	782	56	Nepal	19	88	UAE	69
25	Ecuador	37	57	Netherlands	1672	89	United Kingdom	6326
26	Egypt	706	58	New Zealand	370	90	United States	22718
27	Estonia	102	59	Nigeria	148	91	Uruguay	61
28	Ethiopia	50	60	Norway	354	92	Venezuela	73
29	Ethiopia	50	61	Oman	35	93	Vietnam	127
30	Finland	563	62	Pakistan	384	94	Yemen	1
31	France	3963	63	Peru	37	95	Zimbabwe	16
32	Germany	6292	64	Philippines	63			

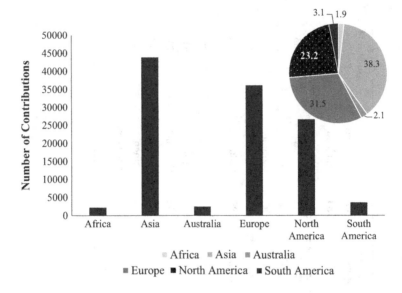

FIGURE 2.3 Geographical-wise distribution of contributions over continents.

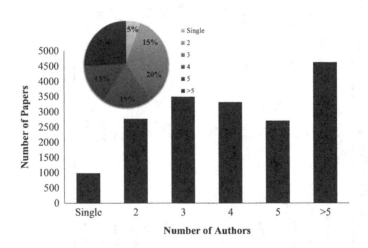

FIGURE 2.4 Authorship pattern of contributions.

papers) with more than five authors. 3,490 papers were studied by three authors, having a 19.5% share. The number of four and five authors collaborations was 18.5% (3,315 papers) and 15.1% (2,705 papers), respectively.

The degree of collaboration is evaluated by Equation 2.1 (Centre, 2014). Results of the degree of collaboration are listed in Table 2.2. The degree of collaboration ranges from 0.88 to 1.00. The average degree of collaboration was calculated as 0.91. After 2009, the average degree of collaboration increases to 0.96 gradually

$$C = \frac{N_m}{N_m + N_s} \qquad (2.1)$$

TABLE 2.2   Degree of Collaboration

Year	Single	Multiple	C
1999	25	83	0.77
2000	25	134	0.84
2001	27	154	0.85
2002	38	168	0.82
2003	26	259	0.91
2004	35	281	0.89
2005	29	375	0.93
2006	40	443	0.92
2007	36	520	0.94
2008	32	569	0.95
2009	60	700	0.92
2010	37	791	0.96
2011	62	969	0.94
2012	52	1114	0.96
2013	68	1310	0.95
2014	59	1448	0.96
2015	67	1690	0.96
2016	73	1764	0.96
2017	91	1996	0.96
2018	90	2149	0.96

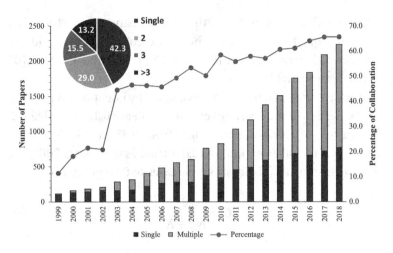

FIGURE 2.5   Year-wise single and collaboration number and percentage of international collaboration.

where C denotes the degree of collaboration and N_m indicated a number of multiple authors. The number of single authors is symbolized by N_s.

In Figure 2.5, the pie chart illustrates the number of international collaborations. 57.7% of published papers were studied by international collaborations. 29% of papers were made by the collaboration of two countries. 42.3% of papers were researched by a single country. However, it should be kept in mind that this high percentage includes collaborations of researchers from the same country where the percentage of single-authored paper was only 6% of the total number of published papers.

Figure 2.5 presents year-wise single and international collaboration as well as the percentage of international collaboration. As is seen in Figure 2.5, the collaboration increases sharply after 2002. After 2007, at least 50% of published papers were studied with collaboration. The presented results in this section reveal and prove the multi-disciplinary nature of green chemistry. The highest collaboration percentage (65.7%) was obtained in 2018.

TABLE 2.3   The Written Language of Papers

Language	Number of Papers
English	17429
Chinese	188
Portuguese	98
Japanese	51
French	28
German	27
Spanish	25
Polish	19
Czech	10
Russian	8
Croatian	2
Finnish	1
Italian	1
Korean	1
Serbo Croatian	1
Turkish	1

The increment of international collaboration (56.5% of published papers in Table 2.3) influenced the language of published papers. 17,429 papers were written in English, as shown in Table 2.3. 2.7% of papers were written in other languages, which are listed in Table 2.3.

2.6 THE NUMBER OF PAGES, CITATIONS, AND REFERENCES DISTRIBUTIONS

Figure 2.6 shows the distribution of a number of pages in each research and review paper. Most research articles' length (8,865 papers) varied from five to ten pages. 3,892 papers had a length between one to fifteen pages, and the number of papers with a length between ten and fifteen pages, was 3,251.

Figure 2.7 presents the number of citations over the years. In the years between 2006 and 2015, the number of citations exceeds 25,000 per year. The number of citations can be described as alive data, which means that these data continue to increase.

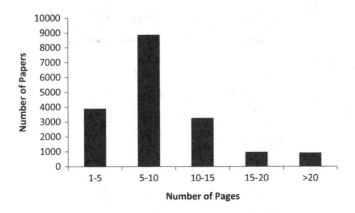

FIGURE 2.6   Number of pages in each paper.

The number of papers versus the number of citation distribution is presented in Figure 2.8. 8,170 research and review papers were cited less than 10 times, which can be expressed as 45.7% percentage. 1,812 papers were not cited. 1,130 papers were cited between 50 and 100 times, a 6.3% percentage. 824 papers were cited more than 100 times; as percentage, this can be defined as 4.6%. High citation quantity can reflect the quality of research.

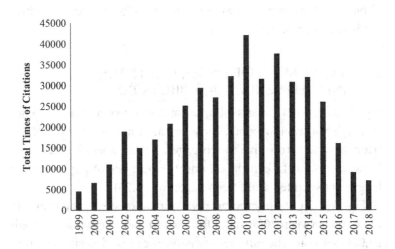

FIGURE 2.7   Year-wise number of citation distribution.

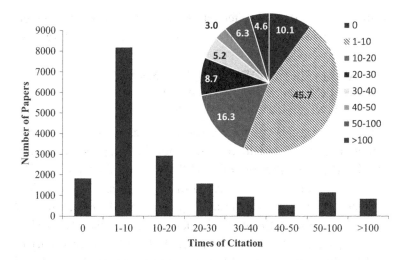

FIGURE 2.8   The number of papers against number of citations.

The histogram of the number of cited references in each paper is shown in Figure 2.9. The histogram of the number of cited references are distributed bimodally since the data was obtained from two different sources: the number of research papers and review papers. As is well known, the review papers were written using a high number of references. The papers were studied and/or written

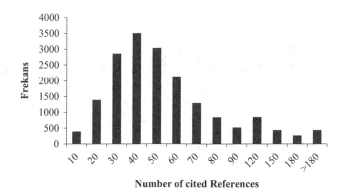

FIGURE 2.9   Histogram of number of cited references in each paper.

using between 30 and 40 cited references approximately on average. The second peak can be seen between 90 and 120 cited references.

2.7 ANALYSIS OF ORGANIZATIONS, JOURNALS, FUNDING, AND FOUNDATION ORGANIZATIONS

Figure 2.10 reveals the top 20 research organizations that contributed to green chemistry. 908 published papers were studied by the contribution of CNRS (France), followed by The Chinese Academy of Sciences with 557 published papers. The Université Côte d'Azur (France) published 507 research and/or review papers. Council of Scientific Industrial Research—CSIR—(India), Islamic Azad University (Iran) and University of California System (United States) contribute with 331, 217, and 272 papers, respectively.

Figure 2.11 illustrates the top 20 journals that published papers related to green chemistry. The number of published papers in the top 20 journals, as shown in Figure 2.11, was 26.2% of the total number of published papers. In other words, one in four papers were printed in these journals. This phenomenon may be because

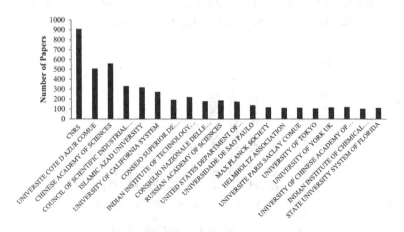

FIGURE 2.10 Top 20 research organizations that contributed to green chemistry.

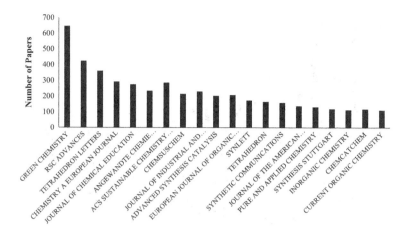

FIGURE 2.11 Top 20 journals, papers published for green chemistry.

these journals reach more readers. *Journal of Green Chemistry* published 646 articles at this time. The *Royal Society of Chemistry Advances* issued 410 papers. The *Tetrahedron Letters, Chemistry: A European Journal,* and *ACS Sustainable Chemistry Engineering* have printed 357, 291, and 281 research and/or review papers, respectively.

Figure 2.12 shows the year-wise funded and non-funded number of papers and the percentage of funding. It is obvious

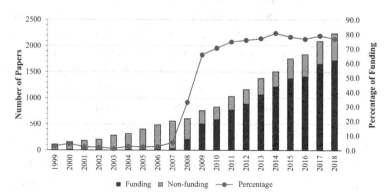

FIGURE 2.12 Year-wise funded and non-funded number of papers and percentage of funding.

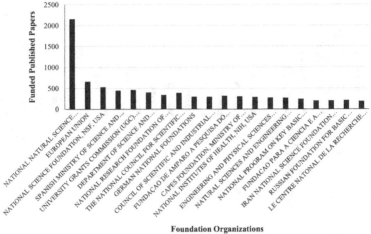

Foundation Organizations

FIGURE 2.13   Top 20 foundation organizations.

that the increment of funding caused an increase in the number of papers. As can be seen in Figure 2.12, the sharp increase was observed after the year 2007. Right after the year 2009, the percentage of funded research papers reached and exceeded the 65% level. The increment of the funded project may increase national and international collaborations and the number of published papers. Nowadays, an increase of analytical equipment cost, consumable prices, and labor cost reveal the importance of project grants. The project grants were supplied by national and international organizations; some of them are mentioned in Figure 2.13.

The top 20 foundation organizations that were acknowledged in published papers are presented in Figure 2.13. The National Natural Science Foundation of China was acknowledged 2,135 times in the published papers; 639 published papers thanked the European Union; The National Science Foundation (NSF) (United States), Spanish Ministry of Science and Innovation, and University Grants Commission (UGC) (India) were mentioned in 509, 426, and 442 papers, respectively. Illustrated top 20 foundation organizations in Figure 2.13 were provided to study 48.2% of the total number of published papers. These results also emphasize the importance of project grants.

REFERENCES

Thavamani, K. *Indian Journal of Animal Research—A Bibliometric Study*, 52 (2014) 1–7. doi:10.5958/0976-2469.2014.01411.0.

Dichiarante, D. Ravelli, A. Albini, Green Chemistry: State of the Art Through an Analysis of the Literature, *Green Chem. Lett. Rev.* 3 (2010) 105–113. doi:10.1080/17518250903583698.

CHAPTER 3

Principles of
Green Chemistry

3.1 PREVENTION OF WASTE

Wastes can be classified as chemical, biological, and radioactive wastes. Waste can involve air contaminants, effluent, hazardous, environmental, and health of life threat substances. Some waste can be used as fertilizer, recovered, and/or recycled. Thus, waste should be distinguished as hazardous or not. Hazardous waste could threaten the environment and health of life (human, plant, and/or animal). Pesticides, heavy metals, mercury, refrigerants are examples of hazardous substances. Hazardous substances effects on health and the environment such as neurological damage and lung cancer may be caused by lead/mercury and asbestos, respectively (SIDA 2017). Figure 3.1 illustrates the percentage of hazardous products for different industries in Sweden (KEMI 2016). Hazardous products in the coke and refined petroleum industry in 2016 were 31.3% of total hazardous commercial products. Export and petrol stations follow with 19.3% and 6.5% of hazardous products, respectively.

Figure 3.2 illustrates the total amount of waste generation in European countries in 2014 ("Eurostat" 2019). The total generation

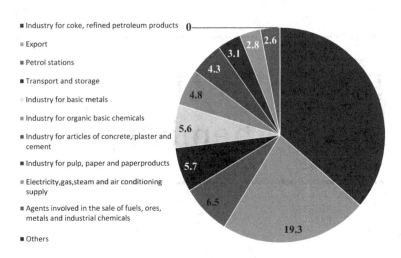

- Industry for coke, refined petroleum products
- Export
- Petrol stations
- Transport and storage
- Industry for basic metals
- Industry for organic basic chemicals
- Industry for articles of concrete, plaster and cement
- Industry for pulp, paper and paperproducts
- Electricity,gas,steam and air conditioning supply
- Agents involved in the sale of fuels, ores, metals and industrial chemicals
- Others

FIGURE 3.1   Percentage of industrial distribution of products.

waste includes hazardous and non-hazardous waste substances such as chemical, medical, health care, and biological wastes; industrial effluent sludge; spent solvents; acid-alkaline or saline wastes; metal waste (ferrous and/or non-ferrous); glass, paper and cardboard; rubber, plastic wastes; and so on. The maximum amount of waste is produced in Germany, with approximately 388

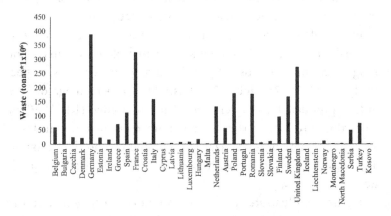

FIGURE 3.2   Hazardous and non-hazardous waste generation in European Countries in 2014.

million tonnes. France and the United Kingdom follow with 325 and 272 million tonnes of waste generation, respectively. At the first blush, the amount of waste generation seems to be relevant with the population of countries; the populations of Germany, France, and the UK, which are crowded countries in Europe, were approximately 83, 67, and 67 million, respectively, in 2017. However, the total populations of Poland (39 million), Bulgaria (7 million), Romania (20 million), and Sweden (10 million) almost equal the population of Germany. But their total amount of generation of waste is 1.8 times that of Germany. The difference comes from the amount of mineral and solidified wastes, including concrete, bricks, and gypsum waste from construction and demolition, insulation materials, mining and quarrying, wastes from flue gas cleaning, etc. (Nuss et al. 2017).

Figure 3.3 shows the total amount of hazardous waste generation in European countries in 2014. The maximum amount of hazardous waste generation is approximately 22 million tonnes and belongs to Germany. Serbia follows Germany with 13.5 million tonnes of hazardous waste generation. Bulgaria, France, Estonia, and Italy generate 12.3, 11, 10.5, and 9 million tonne waste, respectively.

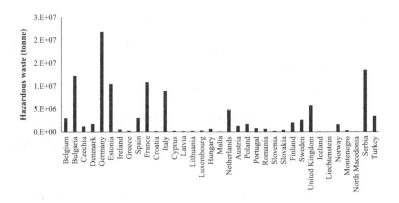

FIGURE 3.3   Total hazardous waste generation in European Countries in 2014.

3.2 PREVENTION OF POLLUTION

The EPA describes the prevention of pollution as any practice that prevents, eliminates, or reduces pollution at all potential and applicable sources such as ore and mineral industries, oil and gas industry, manufacturing, construction, agriculture, paints and solvents, electric power generation utilities, combustion and fires, transportation and mobile equipment etc. (EPA 2018). In 1990, the United States Congress accepted the Pollution Prevention Act, which declares pollution should be recycled, reduced, treated, or prevented in an environmentally safe manner under the responsibility of the EPA. The Prevention Act is done to preserve land, air, water, and other ecosystems. There are many acts to protect the environment for a clean and sustainable future such as:

- Incremental improvement of efficiency in energy consumption.
- Employment of environmentally friendly fuel resources.
- Reducing the use of chemicals and water in agriculture.
- Protection of agricultural areas.
- Improving the process in an industry to generate less waste.
- Replacing toxic chemicals with less/non-toxic ones.
- Recycling and recovering if it's possible.
- Automatically switching off lights, electrical devices when not in use (EPA 2018).
- Use of fewer motor vehicles to reduce emissions.
- Use of environmentally friendly cleaning agents.
- Use of energy-efficient appliances (NPi, Australian Government 2019).

Figure 3.4 shows the air pollution statistics from a report of Environment and Climate Change in Canada in 2017, in which total particulate matter (TPM), SO_x, NO_x, and CO emission in

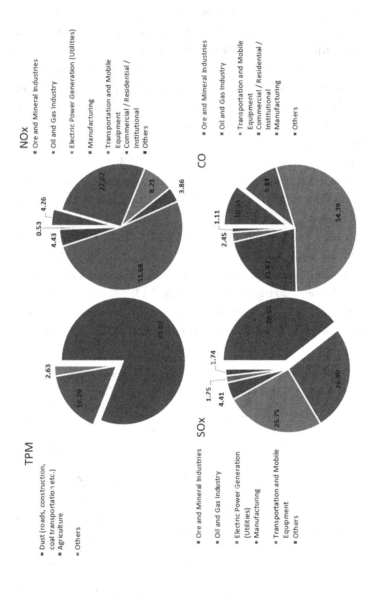

FIGURE 3.4   Air pollutants emissions inventory of Canada in 2017.

the air from different sources was presented. 81% of air pollution from total particulate matter was caused by coal transportation, construction, mine tailings, and paved/unpaved roads. The source of 52% of NO_x pollutant was from transportation and mobile equipment. SO_x emissions have three main sources, which are 39.35% from ore and mineral industries, oil and gas industries with 26.80%, and 25.75% from electric and power generation. CO emissions arise from 54.39% of transportation and mobile equipment and 21.87% of commercial/residential/institutional sectors (NPi 2019).

Figure 3.5 illustrates the air quality of 338 cities in China in 2017. Concentrations of six major pollutants which are $PM_{2.5}$ (particles less than 2.5 μm), PM_{10} (particles less than 10 μm), O_3, SO_2, and NO_2, are given as a percentage (MEE [Ministry of Ecology and Environment] 2017). The World Health Organization (WHO) defined long-term air quality standards that the annual average concentration of $PM_{2.5}$ and PM_{10} should be 10 μg/m³ and 20 μg/m³, respectively. NO_2 annual average concentration should be 40 μg/m³. A 24-hour mean concentration of SO_2 should be 20 μg/m³ and 8 hours mean of O_3 concentration

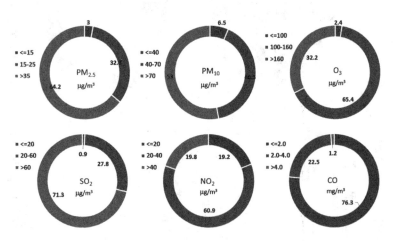

FIGURE 3.5 The air quality of 338 cities of China in 2017.

should be lower than 100 μg/m³ (WHO 2005). The annual average values of air quality of 338 cities in China indicates that all pollutants indicators are higher than the long-term air quality standards of WHO.

Kan et al. (2012) analyzed the air pollutant values of China and concluded that high air pollution may increase mortality or morbidity risks. Moreover, high concentration of air pollutants increased health risks more than in other parts of the world. The importance of health risks can be realized by taking into consideration of China's population (Kan, Chen, and Tong 2012).

The protection and achievement on sustainable remediation of environment and health of life need to devote a large amount of money besides the implementation of the policy. The percentage of expenditure of EU countries for environmental protection services increased from 0.2 to 1.1% of GDP between 2001 and 2009 (Eurostat 2019). Figure 3.6 reveals the monetary value of production in environmental goods and service sectors (EGSS) in the market in 2016 for European countries. In the market, most of the production is sold as well as consumed for wastewater management. The market values of categories

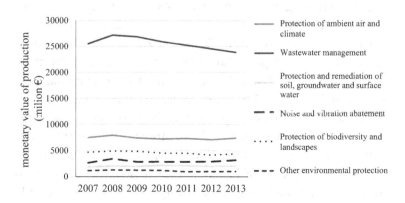

FIGURE 3.6   The market value of EGSS for EU countries.

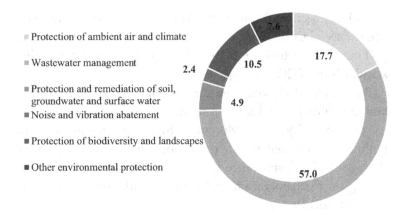

Protection of ambient air and climate

Wastewater management

Protection and remediation of soil, groundwater and surface water

Noise and vibration abatement

Protection of biodiversity and landscapes

Other environmental protection

FIGURE 3.7   Distribution percentage of the market value of categories in EGSS in 2013.

almost remained constant between 2007 and 2013. In Figure 3.7, the percentage of production of wastewater management can be observed better. 57% of the monetary value of production in the market is part of the production of wastewater management in 2013 (Eurostat 2019).

In agriculture, in order to prevent pollution of soil and groundwater, organic farming should be improved. Improvement of organic farming may also affect positively human and animal health. Figure 3.8 illustrates the percentage of change in organic farming area from 2012 to 2016 in European countries. Totally, an organic farming area in European countries is increased by 18.7%. Bulgaria increased the organic farming area from 2012 to 2016 by three times. The organic farming area of Croatia was expanded approximately two times between the same years. France improved the organic farming area by 50%. But they achieve maximum improvement in an organic farming area in the case of hectares. France added 500,000 hectares to the available organic farming area from 2012 to 2016 (Heidorn et al. 2017).

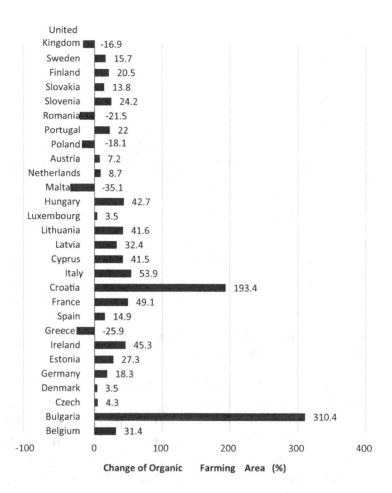

FIGURE 3.8   Percentage of change in the organic farming area between 2012 and 2016.

3.3 PREVENTION OF ACCIDENTS

The cost of handling, treating, and disposing of hazardous chemicals in the United States is evaluated as 4% of manufacturing GDP (Beach, Cui, and Anastas 2009). The prevention of waste has an impact on human health, environment, and cost

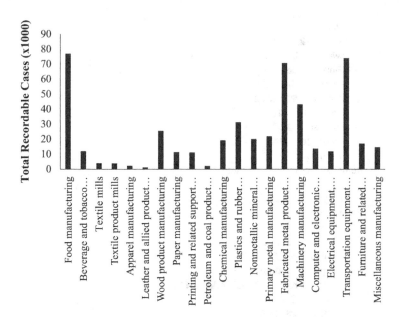

FIGURE 3.9   Numbers of nonfatal occupational injuries and illnesses by case type and ownership 2014.

of manufacturing. The number of nonfatal occupational injuries which was obtained from the Bureau of Labor Statistics, U.S. Department of Labor, in 2014 is illustrated in Figure 3.9. The total number of nonfatal occupational injuries and illness in the manufacturing industry was 483,300 cases. Figure 3.9 shows the diversity of a total number of nonfatal occupational injuries over the selected manufacturing industry. The numbers of nonfatal injuries in food, transportation of equipment, and fabricated metal manufacturing were 76,600, 73,900 and 70,500 respectively.

3.4 USE SAFER SOLVENTS

Solvents are affluently used chemicals. In green chemistry, water, supercritical fluids, ionic liquids, renewable solvents, and liquid polymers are recently used as some of the safer and alternative solvents (Kerton and Marriott 2013) (Anastas and Eghbali 2010). Table 3.1 lists the solvent selection guide into three categories: preferred, usable, and undesirable.

Table 3.2 illustrates the classification of residual solvents used in the pharmaceutical industry. The European Medicines Agency (EMA) distinguishes residual solvents into four categories which are class 1 (should be avoided), class 2 (to be limited), class 3 (low toxic potential), and class 4 (no adequate toxicological data) (EMA report 2017).

Prat et al. (2014) surveyed 51 solvents due to scores and rankings considered by several pharmaceutical companies and institutions, Pfizer, AstraZeneca (AZ), Sanofi-Aventis Research & Development (Sanofi), GlaxoSmithKline Medicines Research Centre (GSK), and ACS Green Chemistry Institute Pharmaceutical Roundtable (GCI-PR). Table 3.3 represents the scores and rankings of 51 solvents of this survey. The companies and institutions considered the solvents in three criteria, safety, health, and environment, and scored each criterion between 1 and 10. The represented scores in Table 3.3 were a summation of the score of each criterion. The overall assessment is indicated by statements which are recommended, problematic, to be confirmed (TBC), banned, substitution advisable (Subst. adv.), substitution required (Subst. req.),

TABLE 3.1   A Green Chemistry-Based Solvent Selection Guide

Category	Solvents
Preferred	water, acetone, ethanol, 2-propanol, ethyl acetate, isopropyl acetate, methanol, methyl ethyl ketone, 1-butanol, t-butanol
Usable	cyclohexane, heptane, toluene, methylcyclohexane, methyl t-butyl ether, isooctane, 2-methyltetrahydrofuran, cyclopentyl methyl ether, xylenes, dimethylsulfoxide, acetic acid, ethylene glycol
Undesirable	pentane, hexane(s), di-isopropyl ether, diethyl ether, dichloromethane, dichloroethane, chloroform, dimethylformamide, n-methylpyrrolidone, pyridine, dimethylacetamide, acetonitrile, tetrahydrofuran, dioxane, Dimethyl ether, benzene, carbon tetrachloride

TABLE 3.2   Classification of Residual Solvents Used in Pharmaceuticals

Category	Solvents
Class 1	Benzene, Carbon tetrachloride, 1,2-Dichloroethane, 1,1Dichloroethene, 1,1,1-Trichloroethane
Class 2	Acetonitrile, Chlorobenzene, Chloroform, Cumene1, Cyclohexane, 1,2-Dichloroethene, Dichloromethane, 1,2-Dimethoxyethane, N,N-Dimetylacetamide, N,N-Dimethylformamide, 1,4-Dioxane, 2-Ethoxyethanol, Ethyleneglycol, Formamide, Hexane, Methanol, 2-Methoxyethanol, Methylbutyl ketone, Methylcyclohexane, N-Methylpyrrolidone, Nitromethane, Pyridine, Sulfolane, Tetrahydrofuran, Tetralin, Toluene, 1,1,2-Trichloroethene, Xylene
Class 3	Acetic acid, Acetone, Anisole, 1-Butanol, 2-Butanol, Butyl acetate, tert-Butylmethyl ether, Dimethyl sulfoxide, Ethanol, Ethyl acetate, Ethyl ether, Ethyl formate, Formic acid, Heptane, Isobutyl acetate, Isopropyl acetate, Methyl acetate, 3-Methyl-1-butanol, Methylethyl ketone, Methylisobutyl ketone, 2-Methyl-1-propanol, Pentane, 1Pentanol, 1-Propanol, 2-Propanol, Propyl acetate
Class 4	1,1-Diethoxypropane, 1,1-Dimethoxymethane, 2,2-Dimethoxypropane, Isooctane, Isopropyl ether, Methylisopropyl ketone, Methyltetrahydrofuran, Petroleum ether, Trichloroacetic acid, Trifluoroacetic acid

hazardous and highly hazardous (HH) (Prat, Hayler, and Wells 2014) (Prat et al. 2015) (Diorazio, Hose, and Adlington 2016).

3.5 REACTION CONDITIONS

The Suzuki cross-coupling reaction has fewer limitations and is environmentally friendly, efficient, and highly useful for the drug industry, compared to Heck reaction, Kumada, Stille, Negishi, and Sonogashiri coupling reactions (Franzén and Xu 2005). The Mannich reaction leads to the production of amino carbonyl compounds, and 1,2-amino alcohol derivatives have been satisfactorily achieved in water (Iwanejko, Wojaczyńska, and Olszewski 2018). Friedel-Crafts alkylation has also another greener process with new approaches that gained the attention of researchers in recent years. The new approach in Friedel-Crafts alkylation reaction allows using a low amount of catalyst. Additionally,

TABLE 3.3 Scores and Rankings of the 51 Solvents

Family	Solvent	AZ	GCI-PR	GSK	Pfizer	Sanofi	Overall
water	water	x	x	24	Preferred	Recommended	Recommended
Alcohols	Methanol	19	14	14	Preferred	Recommended	TBC
	Ethanol	16	13	17	Preferred	Recommended	Recommended
	i-praponol	16	16	17	Preferred	Recommended	Recommended
	n-buthanol	17	13	18	Preferred	Recommended	Recommended
	t-buthanol	20	15	15	Preferred	subst. Adv	TBC
	Benzyl alcohol	x	11	20	x	subst. Adv	TBC
	Ethylene glycol	x	13	21	usable	subst. Adv	TBC
Ketones	Acetone	21	15	15	Preferred	Recommended	TBC
	MEK	21	16	15	Preferred	Recommended	TBC
	MIBK	22	17	15	x	Recommended	TBC
	Cyclohexanone	x	14	20	x	subst. Adv	TBC
Esters	Methyl acetate	x	14	14	x	subst. Adv	TBC
	Ethyl acetate	18	15	16	Preferred	Recommended	Recommended
	i-PrOAc	18	13	18	Preferred	Recommended	Recommended
	n-BuOAc	13	14	21	x	Recommended	Recommended

(Continued)

TABLE 3.3 (CONTINUED) Scores and Rankings of the 51 Solvents

Family	Solvent	AZ	GCI-PR	GSK	Pfizer	Sanofi	Overall
Ethers	Diethyl ether	27	21	3	undesirable	banned	HH
	Diisopropyl ether	x	x	4	undesirable	subst. Adv	Hazardous
	MTBE	24	21	4	usable	subst. Adv	TBC
	THF	23	16	4	usable	subst. Adv	TBC
	Me-THF	24	15	11	usable	Recommended	Problematic
	1,4 Dioxane	28	21	11	undesirable	subst. Req.	Hazardous
	Anisole	18	13	18	x	Recommended	Recommended
	DME	21	23	2	undesirable	subst. Req.	Hazardous
Hydrocarbons	Pentane	x	x	7	undesirable	banned	Hazardous
	Hexane	26	21	1	undesirable	subst. Req.	Hazardous
	Heptane	21	17	14	usable	subst. Adv.	Problematic
	Cyclohexane	25	18	14	usable	subst. Adv.	TBC
	Me-cyclohexane	x	17	16	usable	subst. Adv.	Problematic
	Benzene	x	21	1	undesirable	banned	HH
	Toluene	22	18	11	usable	subst. Adv.	Problematic
	Xylene	19	15	13	usable	subst. Adv.	Problematic
Halogenated	DCM	20	18	5	undesirable	subst. Adv.	TBC
	Chloroform	x	18	4	undesirable	banned	HH
	CCl4	x	19	3	undesirable	banned	HH
	DCE	x	19	4	undesirable	banned	HH
	Chlorobenzene	25	16	18	x	subst. Adv.	Problematic

(Continued)

TABLE 3.3 (CONTINUED) Scores and Rankings of the 51 Solvents

Family	Solvent	AZ	GCI-PR	GSK	Pfizer	Sanofi	Overall
Aprotic polar	Acetronitrile	24	14	14	usable	Recommended	Problematic
	DMF	20	17	7	undesirable	subst. Req.	Hazardous
	DMAc	20	16	4	undesirable	subst. Req.	Hazardous
	NMP	18	16	7	undesirable	subst. Req.	Hazardous
	DMPU	x	x	14	x	subst. Adv.	Problematic
	DMSO	8	15	14	usable	subst. Adv.	Problematic
	Sulfolane	9	13	21	x	subst. Adv.	Recommended
	Nitromethane	x	x	1	x	banned	HH
Miscellaneous	Methoxy ethanol	21	20	8	x	subst. Req.	Hazardous
Acids	Formic acid	20	15	x	x	subst. Req.	TBC
	Acetic acid	17	15	17	usable	subst. Adv.	TBC
	Ac2O	x	16	15	x	subst. Adv.	TBC
Amines	Pyridine	26	16	5	undesirable	subst. Adv.	TBC
	TEA	23	18	3	x	subst. Req.	Hazardous

Prat et al. 2015.

benzyl-, propargyl- and allyl alcohols, or styrenes are started to be used instead of toxic benzyl halides (Rueping and Nachtsheim 2010). Catalytic nanoreactors provide operation of organic reactions in aqueous medium, reducing environmental risks for obtaining chemical sustainability. De Martino et al., (2018) reviewed E-factors of traditional and nanoreactor catalytic reactions in the pharmaceutical industry as shown in Figure 3.10. E-factors of reactions in the pharmaceutical industry decreased significantly using micellar nanoreactors. The advantages of nanoreactors, which are polymersomes, micelles, dendrimers, and nanogels, can be explained as:

- Promotion of cascade reactions

- Generation of hydrophobic materials in aqueous and greener conditions

- Easy recovery of catalyst (De Martino et al. 2018)

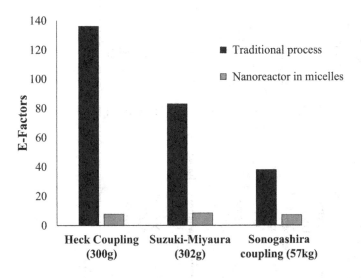

FIGURE 3.10 Comparison of E-factors of the traditional and micellar process. (From De Martino et al. 2018.)

Ultrasound propagates non-hazardous acoustic radiation, having physical and chemical effects on the reaction that formats chemical species not easily achieved under traditional conditions. Ultrasound technology causes cavitation, rapid dispersion, decomposition, and formation of nanostructure, helping to accelerate reactivity of substrates (Cintas and Luche 1999). The advantages of sonochemistry can be summarized as:

- Being an energy-efficient technique

- Potential to improve yields and selectivity

- Potential to increase safety

- Allowing to use of greener medium (aqueous), non-classical reagents

- Providing a possibility to change the course of a reaction (Cintas and Luche 1999)

In recent years, the polymerization reaction has also improved to greener non-hazardous solvents and production with biodegradable polymers due to strict regulations and concerns about waste of the polymer production process and VOCs and hazardous gas. Supercritical fluids (CO_2, etc.) can be used in polymerization reactions. Supercritical fluids having gas-like diffusivities and liquid-like densities can change in solvent density with small fluctuation in temperature or pressure. This type of reaction can be beneficial due to the utilization of inexpensive and non-toxic fluids such as CO_2 (Dubé and Salehpour 2014). High-pressure chemistry (i.e., Q-tube) proposes faster, cheaper, safer, and greener reactions more than conventional techniques. Figure 3.11 presents comparisons of the high-pressure reactor versus microwave (MW) reactor. For all compounds, conversions and yields values of the Q-tube high-pressure reactor are higher than that of the MW reactor. However, both techniques, high-pressure Q-tube

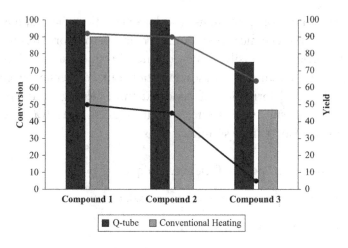

FIGURE 3.11 Conversion and yields for three different compounds at 120°C in 30 min (conversions with bar graph and lines for yields). (From Nacca et al. 2017.)

and microwave reactors, surpass in the manner of reaction time and yield of reaction from conventional heating reactors (Nacca et al. 2017).

Microwave heating in chemistry is used to get green, sustainable, and solvent-free reactions. Advantages of microwave heating in chemistry are:

- Reaction time is decreased tremendously.

- Energy is transferred without contact between source and media.

- Energy transformation can be cut immediately by switching off the device.

- Material selective heating.

- Thermal losses and/or resistance is lower than a traditional heating system.

- Dielectric heating can be occupied for sequential and parallel reactions.

- Energy transfers rather than heat transfer (Demir 2014) (Christopher R. Strauss 2015).

3.6 INCREASE ENERGY EFFICIENCY

Figure 3.12 presents the consumption of primary energy of countries (BPstats 2017). Presented values, million tonnes oil equivalent, involve primary energy comprising commercially traded fuels, including modern renewables used to generate electricity in 2016. Most of the primary energy is consumed by developed and/or developing countries. However, the adoption of green chemistry by developed countries can be observed in Figure 3.13 with

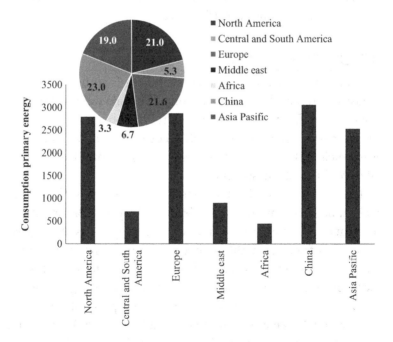

FIGURE 3.12   Consumption of primary energy million tones oil equivalent in 2016.

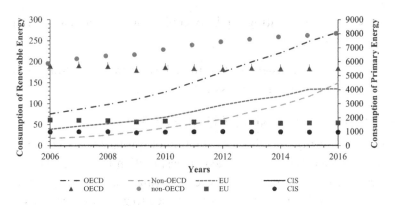

FIGURE 3.13 Consumption of renewable (dots) primary (lines) energy for years (million tons of oil equivalent).

decreasing primary energy consumption values of OECD, EU, and CIS countries over the years (BPstats 2017).

3.7 USE OF RENEWABLE RESOURCES

One of the major goals of green chemistry is to use renewable feedstocks rather than depletable ones. The renewable feedstocks are provided from residues of agricultural products or the wastes of other processes; the depletable feedstocks are obtained from fossil fuels (petroleum, natural gas, or coal) or mining operations (Beach, Cui, and Anastas 2009).

Figure 3.13 illustrates the consumption of renewable energy by economic community countries over the years (BPstats 2017). The data of consumption of renewable energy is based on gross generation from renewable sources including wind, geothermal, solar, biomass, and waste, and does not account for cross-border electricity supply. The consumption of renewable energy has gradually increased in recent years. The consumption of renewable energy for CIS remains almost constant between 2006 and 2016. The primary energy source of members of CIS is petroleum.

Figure 3.14 represents the production of biofuels for 2006 and 2016 (thousand tons of oil equivalent) and increment percentage

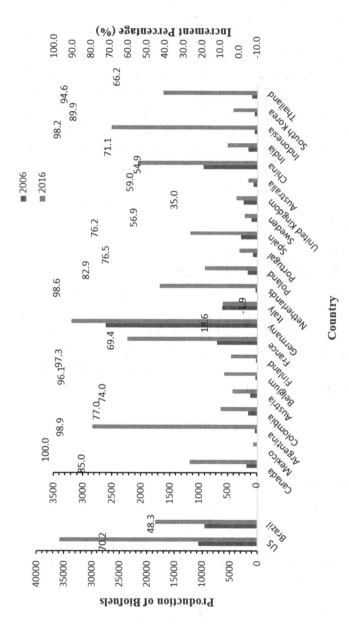

FIGURE 3.14   Production of biofuels for 2006 and 2016 (thousand tons of oil equivalent).

of production from 2006 to 2016 for major biofuel-producing countries (BPstats 2017). The United States and Brazil produced 36,000 and 19,000 tons of oil equivalent biofuels in 2016, respectively. The biofuel production of Germany, Argentina, Indonesia, and China was 3,200, 2,800, 2,500 and 2,000 tons of oil equivalent in 2016, respectively. The minimum increment percentage of biofuel production from 2006 to 2016 was observed as 55%, except for Germany, Italy, and the UK. Since Germany is the third-ranked biofuel producer, the increment percentage of biofuel was below 55%, which represents a relatively low increment or a saturation in the demand. The illustrated data in Figure 3.14 can be interpreted as the countries realized the importance of renewable and sustainable energy resources.

3.8 USE OF CATALYSTS, NOT STOICHIOMETRIC REAGENTS

Catalysts are important in minimizing the waste due to their use in small amounts and ability to carry out a single reaction many times (EPA 2018). Atalay and Ersöz (2016) described nano-catalysts and perovskite-type catalysts, which are the most used types of green catalysts (Atalay and Ersöz 2016). Clark (2001) summarized ideal synthesis as (Clark 2001)

- Atom efficient

- Safe

- One step

- Involving no wasted reagents

- Based on renewable resources

- Environmentally acceptable (including product fate considerations)

Research on green catalysts tends to search novel catalysts (solid acids and bases, ligands, enzymes, biocatalysts, nanoparticles,

TABLE 3.4   The E Factor of Some Chemical Industries

Industry Segment	Product Tonnage[a]	kgwaste [b]/ kgproduct
Oil Refining	$10^6 - 10^8$	<0.1
Bulk chemicals	$10^4 - 10^6$	<1-5
Fine chemicals	$10^2 - 10^4$	5->50
Pharmaceuticals	$10 - 10^3$	25->100

[a] Annual production volume of industry
[b] everything produced except desired product

metal catalyst, etc.), new reaction media (supercritical fluids, ionic liquids, etc.,), and renewable raw materials (green polymers, fatty acid esters, etc.). The E factor gives a valuable measurement for potential level of environmental effects of the chemical process. The E factor can be evaluated as: the mass ratio of waste to the desired product. Increment in E factor indicates an increase in the amount of waste as well as increment of negative environmental effects (Sheldon, Isabel Arends 2007). Table 3.4 shows the E factor of some chemical industries. The highest E factor belongs to the pharmaceutical industry. Contrary to general belief, the E factor of the oil refining industry has the lowest grade, indicating a lower negative environmental impact.

3.9 DESIGN OF DEGRADABLE CHEMICALS AND PRODUCTS

Biodegradation means a breakdown of chemical substances by microorganisms (bacteria, fungi, etc.) to prevent the accumulation of chemical products in the environment (Risotto and Lynn 2018). The Office of Pesticide and Toxic Substances describes the biodegradation rates of chemicals in environments that are affected by chemical variables (i.e., pH, nutrients, toxins, redox potential, and water content), physical variables (temperature, pressure, dilution rate, mixing, sorption, diffusion, and light) and biological variables (microbial interactions, adaptation, etc.) (Lu, Stearns, and Eichenberger 1982). The report concluded that biodegradation processes cannot be described with generic algorithms. Although some effects can be controlled, such as soil moisture, pH, nutrients, oxygen transfer, salinity, toxins, the most important factor—climate—is beyond control (EPA 2018) (Lu, Stearns, and Eichenberger 1982).

Etkin (2010) researched 4,708 ports and harbors of the world to estimate annual oil leakage in marine shipping. Figure 3.15 illustrates the top ten nations' annual lube oil input in ports through stern tube leakage, operational discharges, and overall rate. The results show that every year 28.6 million liters of oil leaks into the sea. In other words, 0.90 liters per second was discharged into the sea and caused an environmental problem (Etkin 2010). The United States Environmental Protection Agency released a report for describing environmentally acceptable lubricant (EAL) which should be demonstrated for biodegradability, low toxicity, and minimum bioaccumulation potential in the aqueous medium compared to conventional lubricants (Albert and Rappoli 2011). Formulations of EAL must consist of 90% biodegradable substance. A maximum 5% of the formulation can be involved in non-biodegradable substances, but they must not be bioaccumulative. The rest of the formulation of EAL should contain readily biodegradable material.

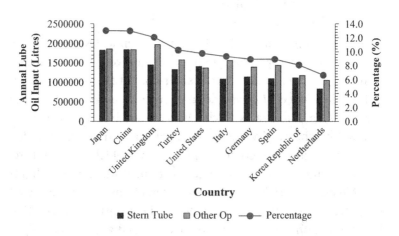

FIGURE 3.15   Top ten nations, evaluated annual lubricant port inputs.

REFERENCES

Albert, R., Rappoli, B., 2011. Environmentally acceptable lubricants. United States Environmental Protection Agency.

Anastas, P., Eghbali, N., 2010. Green chemistry: Principles and practice. *Chem. Soc. Rev.* 39, 301–312. doi:10.1039/b918763b

Atalay, S., Ersoz, G., 2016. *Novel Catalysts in Advanced Oxidation of Organic Pollutants* 7–23. doi:10.1007/978-3-319-28950-2

Banipal, B., Mullins, J., 2003. Aerobic biodegradation of oily wastes. *A Field Guidance Book For Federal On-scene Coordinators*, U.S. Environmental Protection Agency.

Beach, E.S., Cui, Z., Anastas, P.T., 2009. Green chemistry: A design framework for sustainability. *Energy and Environmental Science* 2(10), 1038–49. doi:10.1039/b904997p

BPstats, 2017. BP Statistical Review of World Energy June 2017.

Strauss, Christopher R., Varma, R.S., 2015. Microwaves in green and sustainable chemistry. *Handb. Environ. Chem.* 32, 405–428. doi:10.1007/698_2014_267

Cintas, P., Luche, J.L., 1999. Green chemistry: The sonochemical approach. *Green Chem.* 1, 115–125. doi:10.1039/a900593e

Clark, J.H., 2001. Catalysis for Green Chemistry. *Pure Appl. Chem.* 73, 103–111.

De Martino, M.T., Abdelmohsen, L.K.E.A., Rutjes, F.P.J.T., Van Hest, J.C.M., 2018. Nanoreactors for green catalysis. *Beilstein J. Org. Chem.* 14, 716–733. doi:10.3762/bjoc.14.61

Demir, H., 2014. Experimental study on a novel microwave-assisted adsorption heat pump. *Int. J. Refrig.* 45, 35–43. doi:10.1016/j.ijrefrig.2014.05.001

Diorazio, L.J., Hose, D.R.J., Adlington, N.K., 2016. Toward a more holistic framework for solvent selection. *Org. Process Res. Dev.* 20, 760–773. doi:10.1021/acs.oprd.6b00015

Dubé, M.A., Salehpour, S., 2014. Applying the principles of green chemistry to polymer production technology. *Macromol. React. Eng.* 8, 7–28. doi:10.1002/mren.201300103

EMA, European Medicines Agency. 2017. Committee for human medicinal products ICH guideline Q3C (R7) on impurities: Guideline for residual solvents step 5. *International Conference on Harmonisation (ICH)* 44 (December 2016). www.ema.europa.eu/contact.

EPA, 2018. US environmental protection agency [WWW Document]. web page. URL https://www.epa.gov/greenchemistry/basics-green-chemistry#definition (accessed 11.21.18).

Etkin, D.S., 2010. Worldwide analysis of in-port vessel operational lubricant discharges and leakages. *Proceedings of the 33rd AMOP Technical Seminar on Environmental Contamination and Response*. 7–9 June 2010 Halifax, Ontario, 529–53.

Eurostat [WWW Document], 2019. URL http://appsso.eurostat.ec.euro pa.eu/nui/submitViewTableAction.do (accessed 3.13.19).

Franzén, R., Xu, Y., 2005. Review on green chemistry—Suzuki cross coupling in aqueous media. *Can. J. Chem.* 83, 266–272. doi:10.1139/v05-048

Heidorn, E., Utvik, K., Gengler, C., Alati, K., Collet, D., Attivissimo, V., Colantonio, M., 2017. *Agriculture, Forestry and Fishery Statistics*. 2017 edition. doi:10.2785/570022

Iwanejko, J., Wojaczyńska, E., Olszewski, T.K., 2018. Green chemistry and catalysis in Mannich reaction. *Curr. Opin. Green Sustain. Chem.* 10, 27–34. doi:10.1016/j.cogsc.2018.02.006

Kan, H., Chen, R., Tong, S., 2012. Ambient air pollution, climate change, and population health in China. *Environ. Int.* 42, 10–19. doi:10.1016/j.envint.2011.03.003

KEMI, 2016. Swedish chemical agency [WWW Document]. URL https://www.kemi.se/en/statistics/overview-of-chemicals (accessed 3.13.19).

Kerton F., Marriott R. 2013. Chapter 1: Introduction. In F. Kerton (ed.) *Alternative Solvents for Green Chemistry*, second ed. RSC publishing.

Lu, J.C.S., Stearns, R.J., Eichenberger, B., 1982. Review and critical evaluation of the scientific literature to determine important environmental variables capable of influencing biodegradation rates of chemicals. CalScience Research, Inc.

MEE (Ministry of Ecology and Environment). 2017. Report on the State of the Ecology and Environment in China. http://english.mee.gov.cn/Resources/Reports/soe/SOEE2017/201808/P0201808015 97738742758.pdf.

Nacca, F.G., Merlino, O., Mangiavacchi, F., Krasowska, D., Santi, C., Sancineto, L., 2017. The Q-tube System, A nonconventional technology for green chemistry practitioners. *Curr. Green Chem.* 4, 58–66. doi:10.2174/2213346104666170908160741

NPi, Australian Government, Department of the Environment and Energy, 2019. NPi [WWW Document]. URL http://www.npi.gov.au/reducing-pollution (accessed 3.18.19).

Nuss, P., Blengini, G.A., Haas, W., Mayer, A., Nita, V., Pennington, D., 2017. Development of a Sankey diagram of material flows in the EU economy based on Eurostat Data. doi:10.2760/362116

Prat, D., Hayler, J., Wells, A., 2014. A survey of solvent selection guides. *Green Chem.* 16, 4546–4551. doi:10.1039/c4gc01149j

Prat, D., Wells, A., Hayler, J., Sneddon, H., McElroy, C.R., Abou-Shehada, S., Dunn, P.J., 2015. CHEM21 selection guide of classical- and less classical-solvents. *Green Chem.* 18, 288–296. doi:10.1039/c5gc01008j

Risotto, S., Lynn, A., 2018. American Chemistry Council [WWW Document]. *Environ. Biodegrad. Org. Compd.* URL https://solvents.americanchemistry.com/Biodegradability/

Rueping, M., Nachtsheim, B.J., 2010. A review of new developments in the Friedel–Crafts alkylation – From green chemistry to asymmetric catalysis. *Beilstein J. Org. Chem.* 24, 1–24. doi:10.3762/bjoc.6.6

Sheldon, R.A., 2008. E factors, green chemistry and catalysis: An odyssey. *Chem. Commun.* 3352–3365. doi:10.1039/b803584a

SIDA, 2017. Chemical and hazardous waste, The Swedish International Development Cooperation Agency, Stockholm, Sweden.

U.S. Bureau of Labor Statistics, 2015. Nonfatal occupational injuries and illnesses requiring days away from work, 2015. U.S. Department of Labor. www.bls.gov/iif/oshcdnew.htm (accessed 3.10.19).

U.S. Environmental Protection Agency. 2003. *Aerobic Biodegradation of Oily Wastes: A Field Guidance Book for Feceral On-scene Coordinators.*

CHAPTER 4

Research Trends

4.1 IONIC LIQUID

The number of papers about an ionic liquid topic is presented against years in Figure 4.1. Moreover, the distribution of contribution of countries over the geographic map and collaboration degree are also shown in the same figure. 1,592 papers were published between 1999 and 2018, and 51.8% of papers were released in the last five years. These numbers reveal that the ionic liquid topic has gained the attraction of researchers. The collaboration percentage is also quite high: 55.8%. China made the most contributions, with 457. The United States, India, Iran, and France follow China with 347, 299, 247, and 209 contributions, respectively. The Utilization of ionic liquid in organic chemistry history was started by Paul Walden with the discovery of alkylammonium nitrate [EtNH3][NO3] at the melting point of 12°C during his search for molten salts (Welton, 2018). The utilization of ionic liquids began to spread and the aforementioned statistical data are obtained. Application of ionic liquids can be collated as organic synthesis, pharmaceuticals, cellulose processing, nuclear fuel reprocessing, solar thermal energy, waste recycling, batteries, dispersing agent, tribology, etc. (Handy, 2011) (Kohno and Ohno, 2012).

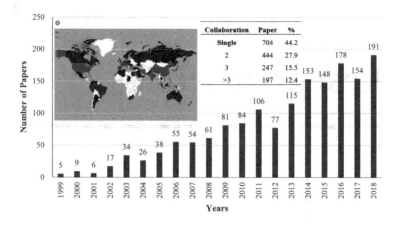

FIGURE 4.1 Collaboration values, number of published papers on the ionic liquid topic and distribution over the geographic map.

The ionic liquids imidazolium and its derivatives were studied in 295 papers. Pyridinium and its derivatives, hexafluorophosphate [PF6} and tetrafluoroborate [BF4], were examined in the papers 86, 101, and 91 times. These ionic liquids were most studied in the research papers.

4.2 SUPERCRITICAL CO_2 FLUID

Supercritical fluid technology has improved since the mid-1980s and uses a fluid state of carbon dioxide where it is at or above its critical point (Sahena et al., 2009). Supercritical fluids are utilized in the extraction and purification processes of materials having low volatility, solubility and/or sensitive to thermal degradation (Díaz-Reinoso et al., 2006). From a green chemistry perspective, supercritical fluid extraction has been boosted in recent years by legal regulations for food and pharmaceutical applications (Díaz-Reinoso et al., 2006). CO_2 as a supercritical fluid is environmentally friendly with non-toxic, non-flammable, and recoverable characteristics against conventional solvents (Sahena et al., 2009). Moreover, supercritical

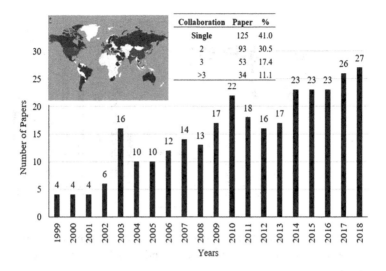

Collaboration	Paper	%
Single	125	41.0
2	93	30.5
3	53	17.4
>3	34	11.1

FIGURE 4.2 Collaboration values, number of published papers on the supercritical CO_2 fluid topic and distribution over the geographic map.

fluids can lead the process faster than conventional solvents, leave almost no trace, and yield a pure residue (Sapkale et al., 2010) (Bhusnure et al., 2015). Figure 4.2 illustrates collaboration values, the number of published papers on the supercritical fluids topic, and distribution over the geographic map. Totally, 305 papers were published between the years 1999 and 2018. The statistical results as shown in Figure 4.2 reveal that the supercritical CO_2 fluids technology is recently modified and promise to spread over the food and pharmaceutical industries widely through becoming cheaper.

4.3 GREEN CATALYSIS

The collaboration percentage, distribution of contribution on world map, and distribution of papers over the years are revealed in Figure 4.3. The bibliometrics analysis presents that 36.4%

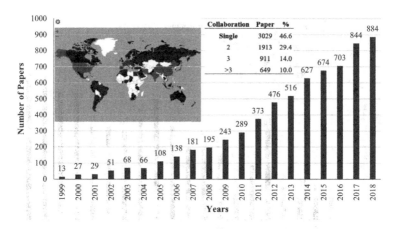

FIGURE 4.3 Distribution of contributions on a geographic map, distribution of papers against years, and collaboration degree for green catalysis.

of 17,889 scientific papers were about catalysis. Moreover, 57.4% of 6,505 papers about a catalyst have been published within the last five years. The statistical data shows the importance and popularity of this topic in recent years. 3,029 papers were studied without national and/or international collaboration. 53.4% of papers were studied with national and/or international collaboration. The highest contributions were made by China with 2,124 and India with 1,506. Iran, the United States, France, and Japan follow, with 1,222, 1,165, 692, and 584 contributions, respectively. The many painted countries in the heat map indicated that the researchers give importance on green catalysis topic all over the world.

Catalysis can be divided into two groups, homogeneous and heterogeneous, according to phase type of catalyst and reactants (Atalay and Ersöz, 2016; Ahluwalia and Kidwai, 2004). As shown in Figure 4.4, homogeneous was cited 283 times and heterogeneous was cited 774 times in published papers. The nano catalyst and/or particles were studied in 1,415 published papers. Biocatalyst, zeolite, titanium oxide, and MOF were encountered as 183, 120, 57, and 41 in published papers, respectively.

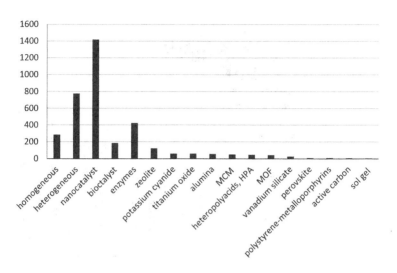

FIGURE 4.4   Types of catalysts.

4.4 POLYMERS

Global production of plastics including HDPE, LDPE, PP, PVC, PET, and PS reached 288 million tons in 2012 (Vargha et al., 2016), and production of plastics continues to increase with increased use of it in many applications in modern society (Mathers and Meier, 2011). Degradation of polymers reflects the loss of its physical and chemical properties by the effects of heat, chemicals, UV lights, microorganisms, etc. Degradation of polymers, especially in the environment, is very important for pollution. For that reason, polymer production is growing very rapidly and having a diverse contribution to green chemistry (Mathers and Meier, 2011).

Figure 4.5 presents the contribution of countries in the production of polymers in the green chemistry perspective and the number of papers against years and collaboration degree. 1,784 papers were studied on topic of polymers. The number of papers increased gradually with increasing years. 39.7% of published papers were made by single institute/department addresses, indicating that researcher/s made the study by oneself or by their selves. The rest

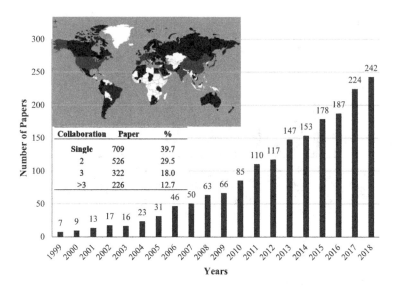

The table embedded in the figure:

Collaboration	Paper	%
Single	709	39.7
2	526	29.5
3	322	18.0
>3	226	12.7

FIGURE 4.5 Number of published papers on the polymers topic and distribution over the geographic map.

of published works, which is 60.3%, were made by national/international collaboration. China and the United States made the most contributions with 568 and 553, respectively, as shown in Figure 4.5 in light grey. India, France, Iran, and Germany made 290, 226, 215, and 204 contributions, respectively.

Gurunathan et al. (2015) classified green polymers according to degradability in environment. According to this classification, abstracts of all papers were investigated, and Figure 4.6 was drawn. Figure 4.6 involves some broad name of green polymeric materials such as biopolymer, polysaccharide, etc. Cellulose, chitosan, starch, and polyaniline were investigated in 86, 68, 39, and 20 published papers.

4.5 MICROWAVE CHEMISTRY

In the last two decades, dielectric heating systems, or microwave heating systems, have begun to be used more frequently than conventional heating systems due to their various advantages, such as:

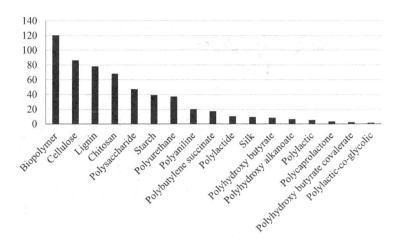

FIGURE 4.6   Types of polymeric materials studied in the published papers.

having high heating rate; providing material selective, non-contact, precise, and controllable heating; transferring energy rather than heat; and providing compact equipment (Haque, 1999) (Pere and Pere, 2002). 6% of published papers (1,077) between 1999 and 2018 about microwave process and distribution of papers over the years are shown in Figure 4.7. Microwave heating system mostly used for drying processes of foods and minerals and regeneration of the adsorbent used for the adsorption of a volatile organic compound (VOC). Idris et al. (2004) investigated the drying of silica sludge under different power conditions (400–1000W) in multimode microwave applicator. The largest of material and lower power input were determined at 1000g and 800W, respectively, according to shorter drying time and higher drying rate. The influence of microwave heating on adsorption selectivity and desorption efficiency of volatile organic compounds on zeolites were researched by Lopez et al. (Alonso Lopez et al., 2007). Their study revealed that microwaves influenced the sorption of polar VOCs more than that of non-polar VOCs. Ania et al. (2005) found that microwave heating increased the phenol adsorption capacity of the activated carbons, in comparison to heating with an electric

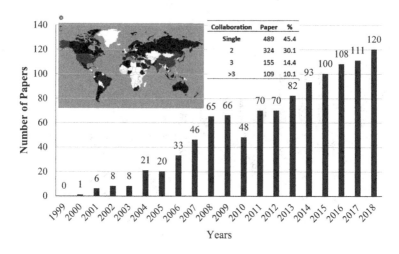

FIGURE 4.7 Number of published papers on the microwave chemistry topic and distribution over geographic map.

furnace. Their conclusion was that microwave heating improves the stability and structure of the activated carbons.

Figure 4.7 illustrates collaboration degree and geographic distribution of published papers. The number of papers increased gradually with increasing years. 409 of 1,077 papers were studied by a single institute/department. 54.6% of the number of papers were worked with national/international collaboration. The highest number of contributions was made by India with 372, as shown in Figure 4.7 in a light grey color. China and the United States made 221 and 181 contributions, respectively.

4.6 PHARMACEUTICALS

Figure 4.8 shows a bibliometric analysis of scientific and reviews papers researched on the pharmaceutical topic between 1999 and 2018. 689 papers were published during these years. 66.2% of papers were published in the last five years. 55.7% of papers were studied with national and international collaborations. The highest number of contributions (125) was made by the United States. India, China, and Brazil have made 94, 69, and 49 contributions, respectively.

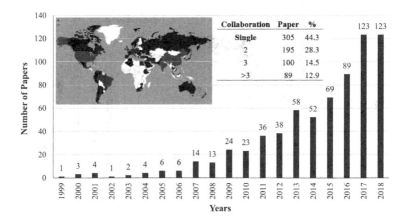

FIGURE 4.8   Number of published papers on the pharmaceutical topic and distribution over the geographic map.

4.7 SONOCHEMICALS

The utilization of toxic chemicals in analytical chemistry increases the environmental and health risk; for that reason, sustainable and greener alternatives have come into prominence. Ultrasound technology is one of the sustainable and non-toxic technologies utilized that have become widespread in recent years. The ultrasound can be classified according to frequency 2–10 MHz, high frequency, and 20–1000 kHz, low-medium frequency. The ultrasound frequency level is above the upper limit of human hearing. The ultrasound can be applied to an elastic medium including water, gas-saturated water, and slurry. The frequency used in industry, chemistry, and nanotechnology is between 20 and 1000 kHz, high-power ultrasound (Sillanpää et al., 2011). The following properties of ultrasound techniques also benefit industrial and/or practical applications.

- Improving extraction production

- Improving non-solvent extraction using aqueous media

- Utilizing alternative greener solvents

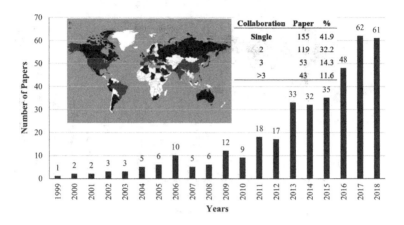

FIGURE 4.9 Number of published papers on the ultrasound topic and distribution over the geographic map.

- Increasing heat transfer rate and provide low-temperature extraction (Tiwari, 2015)

Figure 4.9 illustrates the collaboration degree and geographic distribution of published papers on sonochemical topics. 370 papers were studied on ultrasound topics during the years between 1999 and 2018. 64.3% of papers were published and/or studied after 2013. 58.1% of papers were worked with national/international collaboration. India made the most contributions with 107. China, Brazil, and Iran made 69, 58, and 57 contributions, respectively.

4.8 RENEWABLE RESOURCES

Dramatic consumption of natural resources has raised the importance of renewable and sustainable resources. Renewable resources mean the substance that can be replenished/replaced naturally from its source within the time. Some examples of renewable resources are solar energy, wind energy, geothermal pressure, cultivated plants, air, animals, animal production, water, soil, biomass, biofuels, etc. Non-renewable resources can be explained as coal, oil, nuclear energy, metal, although they

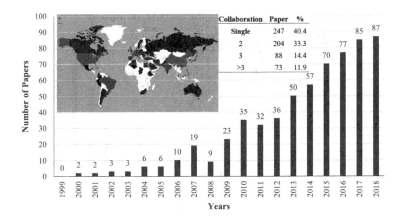

FIGURE 4.10 Distribution of contributions on a geographic map, distribution of papers against years, and collaboration degree for renewable resources.

can be recycled. Figure 4.10 depicts the distribution of contributions on a geographic map, distribution of papers against years, and collaboration degree for renewable resources. 612 of 17,889 papers were on renewable resources. A renewable resource is one of the most important topics of twelve principles of green chemistry. 61.4% of papers on renewable resources were published in the last five years. The United States and China made 197 and 167 contributions, respectively. France, Brazil, and India made 115, 81, and 75 contributions, respectively. The renewable resource topic is also open to collaboration. The statistical data supports that 59.6% of published papers are studied with national/international collaborations.

4.9 NANOTECHNOLOGY

The topic of 3,260 research and review papers, which is 18.2% of whole published papers, were about nanotechnologies. Geographical distribution of published papers is illustrated in Figure 4.11. China made the most contributions with 1285. India, the United States, and Iran followed with 1,076, 836, and 786,

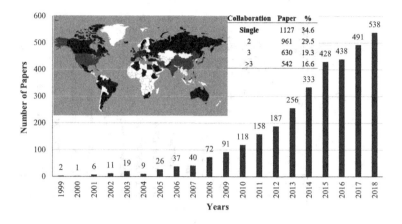

FIGURE 4.11 Distribution of contributions on a geographic map, distribution of papers against years, and collaboration degree for nanotechnology.

respectively. 65.6% of papers were studied with international and national collaboration. The number of published papers increases significantly with increasing years. The year-wise distribution indicates that 68.3% of papers were published over the last five years.

REFERENCES

Ahluwalia, V.K., and M. Kidwai. 2004. Green catalysts. In: *New Trends in Green Chemistry*. Edited by Kidwai, M., Ahluwalia, V.K., Vol. 33. Dordrecht: Springer. doi:10.1097/00005344-199005000-00024.

Alonso Lopez, E., A. Diamy, J. Legrand, and J. Fraissard. 2007. Sorption of volatile organic compounds on zeolites with microwave irradiation. *Studies in Surface Science and Catalysis*. 154: 1866–1871. doi:10.1016/s0167-2991(04)80721-3.

Ania, C. O., J. B. Parra, J. A. Menéndez, and J. J. Pis. 2005. Effect of microwave and conventional regeneration on the microporous and mesoporous network and on the adsorptive capacity of activated carbons. *Microporous and Mesoporous Materials* 85(12): 7–15. doi:10.1016/j.micromeso.2005.06.013.

Atalay, S., and G. Ersöz. 2016. *Novel Catalysts in Advanced Oxidation of Organic Pollutants*. doi:10.1007/978-3-319-28950-2.

Bhusnure, O. G., S. B. Gholve, P. S. Giram, V. S. Borsure, P. P. Jadhav, V. V. Satpute, and Sangshetti, J. N. 2015. Importance of supercritical fluid extraction techniques in pharmaceutical industry: A review. *Indo-American Journal of Pharmaceutical Research* 5(12): 3785–3801.

Díaz-Reinoso, B., A. Moure, H. Domínguez, and J. C. Parajó. 2006. Supercritical CO_2 extraction and purification of compounds with antioxidant activity. *Journal of Agricultural and Food Chemistry* 54(7): 2441–2469. doi:10.1021/jf052858j.

Gurunathan, T., S. Mohanty, and S. K. Nayak. 2015. A review of the recent developments in biocomposites based on natural fibres and their application perspectives. *Composites Part A: Applied Science and Manufacturing* 77(January): 1–25. doi:10.1016/j.compositesa.2015.06.007.

Handy, S. T. 2011. *Ionic Liquids – Classes and Properties Edited*. http://www.issp.ac.ru/ebooks/books/open/Ionic_Liquids_-_Classes_and_Properties.pdf.

Haque, K. E. 1999. Microwave energy for mineral treatment processes—A brief review. *International Journal of Mineral Processing* 57(1): 1–24. doi:10.1016/s0301-7516(99)00009-5.

Idris, A., K. Khalid, and W. Omar. 2004. Drying of silica sludge using microwave heating. *Applied Thermal Engineering* 24(56): 905–918. doi:10.1016/j.applthermaleng.2003.10.001.

Kohno, Y., and H. Ohno. 2012. Ionic liquid/water mixtures: From hostility to conciliation. *Chemical Communications* 48(57): 7119–7130. doi:10.1039/c2cc31638b.

Mathers, R. T., and M. A. R. Meier. 2011. *Green Polymerization Methods: Renewable Starting Materials, Catalysis and Waste Reduction*. Wiley.

Péré, C., and E. Rodier. 2002. Microwave vacuum drying of porous media: Experimental study and qualitative considerations of internal transfers. *Chemical Engineering and Processing* 41(5): 427–436. doi:10.1016/S0255-2701(01)00161-1.

Sahena, F., I. S. M. Zaidul, S. Jinap, A. A. Karim, K. A. Abbas, N. A. N. Norulaini, and A. K. M. Omar. 2009. Application of supercritical CO_2 in lipid extraction—A review. *Journal of Food Engineering* 95(2). Elsevier Ltd: 240–253. doi:10.1016/j.jfoodeng.2009.06.026.

Sapkale, G. N., S. M. Patil, U. S. Surwase, and P. K Bhatbhage. 2010. Supercritical fluid extraction—A review. *International Journal of Chemical Sciences* 8(2): 729–743. https://www.researchgate.net/publication/290484039_Supercritical_fluid_extraction_-_a_review.

Sillanpää, M., T-D. Pham, and R. A. Shrestha. 2011. In S. K. Sharma (ed.) *SpringerBriefs in Molecular Science Green Chemistry for Sustainability.* Edited by Sanjay K Sharma.

Tiwari, B. K. 2015. Trends in analytical chemistry ultrasound: A clean, green extraction technology. *Trends in Analytical Chemistry* 71. Elsevier B.V.: 100–109. doi:10.1016/j.trac.2015.04.013.

Vargha, V., G. Rétháti, T. Heffner, K. Pogácsás, L. Korecz, Z. László, I. Czinkota, L. Tolner, and O. Kelemen. 2016. Behavior of polyethylene films in soil. *Periodica Polytechnica Chemical Engineering* 60(1): 60–68. doi:10.3311/PPch.8281.

Welton, T. 2018. Ionic liquids: A brief history. *Biophysical Reviews* 10(3): 691–706. doi:10.1007/s12551-018-0419-2.

CHAPTER 5

Remarks

THE STUDY COMPREHENDS BIBLIOMETRICS analysis of 17,889 scientific papers indexed in Web of Science from 1999 to 2018. 94 countries contributed to the research and review papers which may also indicate that green chemistry is an interdisciplinary topic.

- The highest contributions were done with 38.3% by countries in Asia and 31.3% by European countries.

- The interdisciplinary of green chemistry was also emphasized by the single-authored papers with minimum percentage, i.e., 5.5%.

- The maximum percentage was observed as 25.9% (4,626 papers) with more than five authors.

- 57.7% of published papers were studied by international collaborations.

- The percentage of funding has a sharp increase after the year 2007. After the year 2009, the percentage of funded research papers reached and exceeded a 65.7% level.

- The National Natural Science Foundation of China was acknowledged 2,135 times in the published papers.

- The Toxics Release Inventory (TRI) Program provides data about environmental releases of toxic and non-toxic chemicals from approximately 22,000 industrial facilities throughout the United States; 1.5 billion kg were released to the air, land, and water in 2015.

- In green chemistry, water, supercritical fluids, ionic liquids, renewable solvents, and liquid polymers are recently being used as safer and alternative solvents.

- 51 solvents, due to scores and rankings, were considered by several pharmaceutical companies and institutions.

- Most of the primary energy is consumed by developed and/or developing countries. However, the adoption of green chemistry by developed countries can be observed in Figure 3.13 with decreasing primary energy consumption values of OECD, EU, and CIS countries over the years.

- The consumption of renewable energy has gradually increased in recent years.

- The United States and Brazil produced 36,000 and 19,000 tons of oil equivalent biofuels in 2016, respectively.

- The United States Environmental Protection Agency released a report describing environmentally acceptable lubricant (EAL), which should be demonstrated for biodegradability, low toxicity, and minimum bioaccumulation potential in the aqueous medium compared to conventional lubricants.

- 0.90 liters lubricants per second were discharged into the sea and caused environmental problems. Formulations of EAL must consist of 90% biodegradable substance.

- 17,889 research and review papers were investigated according to sub-topics such as ionic liquids, supercritical CO_2 fluid, catalysis, polymers, microwave technology, sonochemical technology, nanotechnology, pharmaceutical, and renewable resources. Catalysis and nanotechnology topics are the most popular topics, encountered in 6,505 and 3,260 papers, respectively.

Index

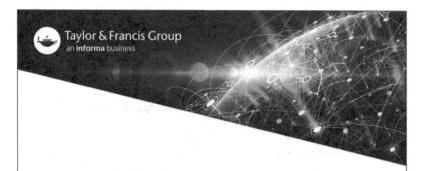

9781032337586